Introduction to the Basics of Real Analysis

Introduction to the Basics of Real Analysis

Harendra Singh

Post-Graduate College Ghazipur, India

H M Srivastava

University of Victoria, Canada

NEW JERSEY · LONDON · SINGAPORE · BEIJING · SHANGHAI · HONG KONG · TAIPEI · CHENNAI · TOKYO

Published by

World Scientific Publishing Co. Pte. Ltd.

5 Toh Tuck Link, Singapore 596224

USA office: 27 Warren Street, Suite 401-402, Hackensack, NJ 07601

UK office: 57 Shelton Street, Covent Garden, London WC2H 9HE

Library of Congress Cataloging-in-Publication Data
Names: Singh, Harendra, author. | Srivastava, H. M., author.
Title: Introduction to the basics of real analysis / Harendra Singh,
    Ravindrapuri Ghazipur, India, H.M. Srivastava, University of Victoria, Canada.
Description: Hackensack, NJ : World Scientific, [2024] |
    Includes bibliographical references and index.
Identifiers: LCCN 2023032832 | ISBN 9789811278211 (hardcover) |
    ISBN 9789811278228 (ebook for institutions) | ISBN 9789811278235 (ebook for indidiuals)
Subjects: LCSH: Functions of real variables--Textbooks. | Functional analysis--Textbooks. |
    Numbers, Real--Textbooks.
Classification: LCC QA331.5 .S533 2024 | DDC 515/.8--dc23/eng/20231011
LC record available at https://lccn.loc.gov/2023032832

British Library Cataloguing-in-Publication Data
A catalogue record for this book is available from the British Library.

For any available supplementary material, please visit
https://www.worldscientific.com/worldscibooks/10.1142/13467#t=suppl

Desk Editors: Sanjay Varadharajan/Rok Ting Tan

Typeset by Stallion Press
Email: enquiries@stallionpress.com

Printed in Singapore

Dedicated to
My beloved Parents

Preface

This book presents an introduction to the key topics in Real Analysis and makes the subject easily understandable for the learners. The book is primarily useful for students of mathematics and engineering studying the subject of Real Analysis. It includes many examples and exercises at the end of chapters. This book is very authentic for students, instructors, as well as those doing research in areas demanding a basic knowledge of Real Analysis. It describes several useful topics in Real Analysis such as sets and functions, completeness, ordered fields, neighborhoods, limit points of a set, open sets, closed sets, countable and uncountable sets, sequences of real numbers, limit, continuity and differentiability of real functions, uniform continuity, point-wise and uniform convergence of sequences and series of real functions, Riemann integration, improper integrals, and metric spaces.

This book consists of 12 chapters and it is organized as follows:

In the initial chapter, we present the background needed for the study of Real Analysis. First, we discuss sets and their basic properties and then functions, their types, and some of their properties. We understand that most of the students are familiar with the properties of sets and functions, so here we give only a brief discussion about these topics.

Chapter 2 presents three basic properties of the real number system. First we discuss field axioms using two operations: addition (+) and multiplication (.). Then we give order properties which are useful to solve inequalities in Real Analysis. Further, we define absolute value by the use of order property. In continuation, we provide definitions of bounded set, supremum, and infimum. In the next section, we present the most important property of real number systems known as the completeness axiom and, by using this, the Archimedean property is proved.

Chapter 3 is important and useful for the study of the next chapters. In this chapter, first we define the neighborhoods of a point. We then discuss the interior points and open sets, and we finally describe some basic theorems related to these. Afterward, we give the definition of limit points, derived sets, adherent points, closed sets, dense sets, and perfect sets. We deal briefly with the properties of closed sets. Finally, we discuss countable and uncountable sets and provide some important examples of these sets.

Chapter 4 is devoted to a special class of functions whose domain is the set of natural numbers and co-domain is a subset of real numbers. Mainly, we focus on real sequences. First, we introduce the meaning of the convergence of real sequences and then present some basic results which are useful for the convergence of sequences. Further, we discuss some important theorems related to the convergence of sequences such as the Sandwich Theorem, the Monotone Convergence Theorem, the Cauchy First and Second Theorems, and the Ceasaro Theorem. In continuation, we introduce the Cauchy sequences and present some important results for the Cauchy sequences. Lastly, we present subsequences of real numbers.

In Chapter 5, we extend our concept of limit for real functions and study the most important class of functions that arises in Real Analysis: The class of continuous functions. First, we define the continuity of functions at a point and on a set. We then define the discontinuity of functions and their types. Next, we present the algebra of continuous functions. We extend our study and describe the concept of sequential continuity, and then use this concept to discuss the continuity of Dirichlet's function. We also establish some important properties of continuous functions, such as the fact that continuous functions on closed and bounded intervals must attain their minimum and maximum values, followed by the intermediate value property and fixed point property.

In Chapter 6, we define the notion of uniform continuity of real functions, observing that the continuity of a function is local in character whereas the uniform continuity is global in character.

In the present chapter, we discuss uniform continuity and provide some examples of uniform and non-uniform continuous functions. Further, we show that, on a closed bounded interval, every continuous function is also uniformly continuous.

Chapter 7 starts with the notion of derivative of a real function. We are already familiar with the geometrical concept of derivative. Therefore, in the present chapter, we discuss the mathematical concept of the derivative. First we introduce the definition and some examples of derivatives. We then discuss the algebra of derivatives and the meaning of the sign of derivatives. Lastly, we study some important theorems on derivatives such as Darboux's Theorem, Intermediate Value Theorem, Rolle's Theorem, Langrage's Mean Value Theorem, and Cauchy's Mean Value Theorem.

In Chapter 8, we extend our concepts of sequences and consider the sequences of real functions rather than real numbers as discussed in Chapter 4. It starts with two notions of convergence of sequences of functions: point-wise convergence and uniform convergence. We discuss some results for the test of convergence of sequences of functions. In the next section, we define the notion of point-wise and uniform convergence of a series of functions. In the end, we present some important results such as the Weierstrass M-test, the Abel test, and the Dirichlet test for the uniform convergence of a series of functions.

In Chapter 9, we extend our notion of limits and continuity for the real functions of two variables. First, we introduce some definitions and then define limits and continuity of functions of two variables. Further, we present Taylor's theorem for the functions of two variables. In the next section, we give a basic idea for the maxima and minima of functions of two variables. Finally, we furnish Lagrange's method of undetermined multipliers for the maxima and minima of functions of three variables.

In Chapter 10, we first give the definition of the Riemann integration of a bounded function on closed bounded intervals and then discuss some important theorems on Riemann integration. We also establish Riemann integrability of several important classes of functions: step functions, continuous functions, and monotonic functions. Lastly, we present the first and second forms of the mean value theorem and the fundamental theorem of calculus. A number of important consequences of these theorems are also given.

Chapter 11 presents the types of improper integrals and some important theorems which include comparison test, μ-test, Abel's test, and Dirichlet's test for their convergence. We also discuss the absolute convergence of improper integrals of the first kind.

Chapter 12 deals with metric spaces. First, we discuss the concept and properties of metric spaces. Further, we investigate open and closed sets in a metric space. In continuation, we present convergence and Cauchy sequences in a metric space. In the next section, we discuss the completeness of metric spaces. Finally, we present Cantor's intersection theorem.

About the Authors

Harendra Singh is an Assistant Professor in the Department of Mathematics, Post-Graduate College, Ghazipur-233001, Uttar Pradesh, India. He did his Master of Science (M.Sc.) in Mathematics from Banaras Hindu University, Varanasi, and Ph.D. in Mathematics from Indian Institute of Technology (BHU), Varanasi, India. He is qualified for GATE, JRF, and NBHM in Mathematics. He is also awarded post-doctoral fellowship (PDF) in Mathematics from National Institute of Science Education and Research (NISER), Bhubaneswar, Odisha, India. He primarily teaches subjects like real and complex analysis, functional analysis, Abstract Algebra, and measure theory in post-graduate level course in mathematics. **He was listed in the top 2% scientist list published by Stanford University.** His areas of interest are Mathematical Modelling, Fractional Differential Equations, Integral Equations, Calculus of Variations, and Analytical and Numerical Methods. His 50 research papers have been published in various journals of repute with an h-index of 25 and an i10 index of 43. He has edited **seven books** out of which four are from Taylor and Francis, two from Springer, and one from Elsevier. He has delivered lectures in a number of national and international conferences and workshops. He is the editor and reviewer of various journals.

 **H. M. Srivastava** is a Professor Emeritus in the Department of Mathematics and Statistics, University of Victoria, British Columbia V8W 3R4, Canada. He earned his Ph.D. degree in 1965 while he was a full-time member of the teaching faculty at the Jai Narain Vyas University of Jodhpur in India (since 1963). Professor Srivastava has held (and continues to hold) numerous Visiting, Honorary, and Chair Professorships at many universities and research institutes in different parts of the world. Having received several D.Sc. (honoris causa) degrees as well as honorary memberships and fellowships of many scientific academies and scientific societies around the world, he is also actively associated editorially with numerous international scientific research journals as an Honorary or Advisory Editor or as an Editorial Board Member. He has also edited many Special Issues of scientific research journals as the Lead or Joint Guest Editor, including, for example, the MDPI journals, *Axioms, Mathematics,* and *Symmetry,* the Elsevier journals, *Journal of Computational and Applied Mathematics, Applied Mathematics and Computation, Chaos, Solitons & Fractals, Alexandria Engineering Journal,* and *Journal of King Saud University — Science,* the Wiley journal, *Mathematical Methods in the Applied Sciences,* the Springer journals, *Advances in Difference Equations, Journal of Inequalities and Applications, Fixed Point Theory and Applications,* and *Boundary Value Problems,* the American Institute of Physics journal, *Chaos: An Interdisciplinary Journal of Nonlinear Science,* the American Institute of Mathematical Sciences journal, *AIMS Mathematics,* the Hindawi journals, *Advances in Mathematical Physics, International Journal of Mathematics and Mathematical Sciences,* and *Abstract and Applied Analysis,* the De Gruyter (now the Tbilisi Centre for Mathematical Sciences) journal, *Tbilisi Mathematical Journal,* the Yokohama Publisher journal, *Journal of Nonlinear and Convex Analysis,* the University of Nis journal, *Filomat,* the Ministry of Communications and High Technologies (Republic of Azerbaijan) journal, *Applied and Computational Mathematics: An International Journal,* and so on. He is a Clarivate Analytics [Thomson Reuters] (Web of Science) Highly Cited Researcher. Professor Srivastava's research interests include several areas of pure

and applied mathematical sciences, such as, for example, real and complex analysis, fractional calculus and its applications, integral equations and transforms, higher transcendental functions and their applications, q-series and q-polynomials, analytic number theory, analytic and geometric inequalities, probability and statistics, and inventory modeling and optimization. He has published 36 books, monographs, and edited volumes, 36 book (and encyclopedia) chapters, 48 papers in international conference proceedings, and more than 1350 peer-reviewed international scientific research journal articles, as well as forewords and prefaces to many books and journals. His research papers have been published in various journals of repute with an h-index of 93 and an i10-index of 714.

Acknowledgements

It is my immense pleasure and proud privilege to express my sincere thanks to **Professor H. M. Srivastava** who is a co-author of this book for his support in this work. Without his keen supervision and sagacious guidance it would not have been possible for me to complete the work.

I pay my respects and heartiest thanks to my **father, mother, elder brother and his wife, my nephew Raghav Singh**, my best friends and students who encouraged me to write this book-without their blessings and affection this book would not have been complete. I also extend my special thanks to my wife **Ms. Priti Singh** and my **son Madhav Singh** for their deepest love, endless patience and continued support, shown during the whole tenure of my work.

Above all, I bow down before the **Almighty** who has made everything possible.

Contents

Chapter 1

Sets and Functions

Two basic tools of real analysis are sets and functions. In this chapter, first we discuss sets and their basic properties and then functions, their types, and some of their properties. Since we have already studied sets and functions in the intermediate (the junior college) level, here we give only a brief discussion about these topics.

1.1 Sets

The word set is a synonym of "family", "collection" or "class". We can say that the set is a collection of objects which are "well-defined". The word "well-defined" means all the members of the set have common properties and the repetition of the elements will not take place. The members of a set may also be called the elements of the set. Capital letters like A, B, C, etc. are used to name a set and the members of the set are written in small letters like a, b, c, etc. If we write $a \in A$, it means that "a is a member of A" or "a belongs to A". Also, if $a \notin A$, then "a is not a member of A" or "a does not belong to A".

Some important sets are as follows:

N: The set of natural numbers;
Z: The set of integers;
Q: The set of rational numbers;
R: The set of real numbers;
C: The set of complex numbers.

We can represent any set in the following two standard ways.

1. By writing all the elements of the set.

    I. If a set A has first five natural numbers, then we represent this set as $A = \{1, 2, 3, 4, 5\}$.

    II. If a set B is a collection of vowels in English alphabets, then we represent this set as $B = \{a, e, i, o, u\}$.

2. By means of a property which is common to all its elements. This type of sets are usually written as $\{x : P(x)\}$, where $P(x)$ is a property possessed by every element of the set.

    I. The set A of all integers can be written as follows:

    $$A = \{x : x \in \mathbf{Z}\}.$$

    II. The set X of even natural numbers can be written as

    $$X = \{2n : n \in \mathbf{N}\}.$$

Example. The set

$$Y = \left\{ x \in \mathbf{Z} : x^2 - \frac{3}{2}x + \frac{1}{2} = 0 \right\}$$

consists of those integers which satisfy the indicated equation. The solutions for the indicated equations are $= 1, \frac{1}{2}$. But $\frac{1}{2}$ is not an integer. So, the set Y can simply be written as $Y = \{1\}$.

Null set. A set having no element is a null set or a set in which no element satisfies the describing property. It is also called **empty set** or **void set**. This type of set is denoted by Φ or $\{\ \}$.

    I. $\Phi = \{n : n \text{ is a natural number lying between 1 and 2}\}$.

    II. $\Phi = \{x \in \mathbf{Z} : x^2 - 7x + 12 = 0 \text{ and } x < 2\}$. The solution of the indicated equation is $x = 3, 4$. Both are integers, but none of them satisfies the property $x < 2$. So the given set is an empty set.

Subsets: If A and B are two sets, then we say that A is a subset of B if each member of the set A is also a member of the set B. This means if $x \in A \Rightarrow x \in B$. It is denoted as $A \subseteq B$. Some important points about the subsets are as follows:

 I. If $A \subseteq B$, then it can also read as B is a super set of A.
 II. If $A \subset B$, then it means that $\exists$ (there exist) some $b \in B$ such that $b \notin A$. It is read as A is a proper subset of B.
 III. Φ and A are two subsets of any set A, that is, $\Phi \subset A$ and $A \subseteq A$.

Example. $\mathbf{N} \subseteq \mathbf{Z} \subseteq \mathbf{Q} \subseteq \mathbf{R} \subseteq \mathbf{C}$.

Equality of sets: If A and B are two sets, then we say that set A is equal to set B if they have the same elements. It is written as $A = B$. If we want to show that two sets A and B are equal, then we need to show that

$$A \subseteq B \quad \text{and} \quad B \subseteq A.$$

Universal set. In any discussion of sets, if all sets are subsets of a set, then that set (whose subsets are all sets) is called a universal set. We can treat $\mathbf{R}, \mathbf{Z}, \mathbf{N}$, and $\mathbf{Q}$ as universal sets.

1.2   Set Operations

Union. The union of two sets A and B is denoted as $A \cup B$, and this set contains all those elements which are in A or in B or in both A and B:

$$A \cup B = \{x : x \in A \,\text{or}\, x \in B\}.$$

Intersection. The intersection of two sets A and B is denoted as $A \cap B$, and this set contains all those elements which are in both A and B:

$$A \cap B = \{x : x \in A \text{ and } x \in B\}.$$

If $A \cap B = \Phi$, then the sets A and B are called disjoint sets.

Complement. The complement of a set A is denoted as A' or A^c and this set contains all those elements which are not in A:

$$A' = \{x : x \notin A\}$$

or

$$A' = \{x \in X : x \notin A\}, \text{ where } X \text{ is a universal set, i.e. } A \subset X.$$

Difference. The difference of two sets A and B is denoted as $A - B$, and this set contains all those elements which are in A but not in B:

$$A - B = \{x : x \in A \text{ and } x \notin B\}.$$

Symmetric difference. The symmetric difference of two sets A and B is denoted as $A \Delta B$, and this set contains all those elements which are in A or in B, but not in both A and B:

$$A \Delta B = (A \cup B) - (A \cap B). \tag{1.1}$$

Example. Let $X = \{1, 2, 3, \ldots, 10\}$, $A = \{1, 2, 3, 4, 5\}$ and $B = \{4, 5, 6, 7, 8, 9\}$. Then

$$A \cup B = \{1, 2, 3, \ldots, 9\},$$
$$A \cap B = \{4, 5\},$$
$$A' = \{6, 7, 8, 9, 10\},$$
$$A - B = \{1, 2, 3\},$$
$$A \Delta B = \{1, 2, 3, 6, 7, 8, 9\}.$$

In the above discussion, we have discussed the union and intersection of two sets. In a similar way, we can define the union and intersection of arbitrary collections of sets. Arbitrary collections can be finite or infinite. If we have collections $\{A_1, A_2, \ldots, A_n\}$ of sets, then their union and intersection are as follows:

$$\bigcup_{i=1}^{n} A_i = A_1 \cup A_2 \cup \cdots \cup A_n = \{x : x \in A_i \text{ for some } i = 1, 2, \ldots, n\},$$

$$\bigcap_{i=1}^{n} A_i = A_1 \cap A_2 \cap \cdots \cap A_n = \{x : x \in A_i \ \forall i = 1, 2, \ldots, n\}.$$

If we have infinite collections $\{A_1, A_2, A_3, \ldots\}$ of sets, then their union and intersection are as follows:

$$\bigcup_{i=1}^{\infty} A_i = \{x : x \in A_i \text{ for some } i = 1, 2, \ldots\},$$

$$\bigcap_{i=1}^{\infty} A_i = A_1 \cap A_2 \cap \cdots = \{x : x \in A_i \ \forall i = 1, 2, \ldots\}.$$

Arbitrary union. If we write $\cup_{i \in \Delta} A_i$ then there are three possibilities of union, as follows:

1. If $\Delta = \{1, 2, \ldots, n\}$, then we write $\cup_{i=1}^{n} A_i$, that is, finite union of sets.
2. If $\Delta = \{1, 2, \ldots, n, \ldots\}$ which means indexing is from countable set, then we write $\cup_{i=1}^{\infty} A_i$, that is, countable union of sets.
3. If indexing is from some uncountable set, then $\cup_{i \in \Delta} A_i$ is called uncountable union of sets.

Arbitrary union means union that can be from any of above three categories. Similarly, we can define arbitrary intersection.

1.3   Algebra of Sets

i. **Commutative laws:** For any two sets A and B, we have

$$A \cup B = B \cup A, \quad A \cap B = B \cap A.$$

ii. **Associative laws:** For any three sets A, B, and C, we have

$$A \cup (B \cup C) = (A \cup B) \cup C, \quad A \cap (B \cap C) = (A \cap B) \cap C.$$

iii. **Distributive laws:** For any three sets A, B, and C, we have

$$A \cap (B \cup C) = (A \cap B) \cup (A \cap C), \quad A \cup (B \cap C) = (A \cup B) \cap (A \cup C).$$

iv. **De Morgan's laws:** For any two sets A and B, we have

$$(A \cup B)' = A' \cap B', \quad (A \cap B)' = A' \cup B'.$$

If we have collections $\{A_1, A_2, \ldots, A_n\}$ of sets, then De Morgan's law is given by

$$\left(\bigcup_{i=1}^{n} A_i \right)' = \bigcap_{i=1}^{n} A_i'.$$

1.4   Functions

Product of sets. If A and B are two non-empty sets, then their Cartesian product is a collections of all ordered pairs (a, b), where $a \in A$ and $b \in B$. It is written as follows:

$$A \times B = \{(a, b) : a \in A \text{ and } b \in B\}.$$

Example. If $A = \{1, 2, 3\}$ and $B = \{1, 4\}$, then

$$A \times B = \{(1, 1), (1, 4), (2, 1), (2, 4), (3, 1), (3, 4)\}.$$
$$B \times A = \{(1, 1), (4, 1), (1, 2), (4, 2), (1, 3), (4, 3)\}.$$

Example. If $A = \mathbf{R}$ and $B = \mathbf{R}$, then $\mathbf{R} \times \mathbf{R}$ is the collection of all points from the Cartesian plane and is given by

$$\mathbf{R}^2 = \mathbf{R} \times \mathbf{R} = \{(a, b) : a \in R \text{ and } b \in R\}.$$

If we have the collection $\{A_1, A_2, \dots, A_n\}$ of non-empty sets, then

$$A_1 \times A_2 \times \cdots \times A_n = \{(a_1, a_2, \dots, a_n) : a_i \in A_i \; \forall i = 1, 2, \dots, n\}.$$

Relations. If A and B are two non-empty sets, then any non-empty subset of $A \times B$ forms a relation. A function is a special type of relation in which each element of A is uniquely related to elements of B. There are two basic properties of a function:

1. Function (f) is defined for each member of A.
2. Each member of A has a unique image in B.

It means that, for each $a \in A$, there exist a unique $b \in B$ such that $(a, b) \in f$, that is, if $(a, b) \in f$ and $(a, b') \in f$, then $b = b'$. We can also write $f : A \to B$ to mean that f is a function from A to B.

Domain and range. The set of the first element in a function is a domain and it is denoted by $D(f)$. The collection of image points of the element of A is a range and it is denoted by $R(f)$. It means that $D(f) = A$ and $R(f) \subseteq B$.

Example 1. Let $A = \{1, 2, 3\}$ and $B = \{4, 5, 6\}$. Define $f : A \to B$ by

a. $f(1) = 4, f(2) = 5$, and $f(3) = 6$. Then f is a function because it follows both properties of a function.
b. $f(1) = 4, f(1) = 5, f(2) = 5$, and $f(3) = 6$. Then f is not a function because it follows only the first property. $1 \in A$ does not have a unique image. It has two images 4 and 5.
c. $f(1) = 4$ and $f(2) = 5$. Then f is not a function because $3 \in A$, but f is not defined at this point. So the first property is violated.

Example 2. Let $A = \{1, 2, 3\}$ and $B = \{1, 3, 5, 7, 9\}$. Define $f : A \to B$ by
$f(1) = 1, f(2) = 5$, and $f(3) = 9$. Then f is a function and $D(f) = A$ and $R(f) = \{1, 5, 9\}$.

Example 3. Let $A = \{a \in \mathbf{R} : a \neq 2\}$ and define $f : A \to B$ by $f(a) = \frac{a}{a-2} \forall a \in A$. Now, we will calculate the range of f. Let $b \in B$ such that $b = \frac{a}{a-2}$. Then we will calculate a in terms of b. On solving $a = \frac{2b}{b-1}$, which is meaningless if $b = 1$. So $R(f) = \mathbf{R} - \{1\}$.

One–one function. A function $f : A \to B$ is said to be one–one if, whenever $a \neq b$, then $f(a) \neq f(b)$, that is, two different elements cannot have the same image. In other words, if $f(a) = f(b)$, then $a = b$.

Onto function. A function $f : A \to B$ is said to be onto if each element of B has a pre-image, that is, $R(f) = B$.

Example 1. Let $A = \{1, 2, 3\}$ and $B = \{4, 5, 6\}$. Define $f : A \to B$ by $f(1) = 4, f(2) = 5$, and $f(3) = 6$. Then f is one–one and onto.

Example 2. Let $A = \{1, 2, 3\}$ and $B = \{4, 5, 6, 7\}$. Define $f : A \to B$ by $f(1) = 4, f(2) = 5$, and $f(3) = 6$. Then f is one–one but not onto because $R(f) = \{4, 5, 6\} \neq B$.

Example 3. Let $A = \{1, 2, 3\}$ and $B = \{4, 5\}$. Define $f : A \to B$ by $f(1) = 4, f(2) = 5$, and $f(3) = 5$. Then f is onto, but not one–one, because $2 \neq 3$ but $f(2) = f(3)$.

Example 4. Let $A = \{a \in \mathbf{R} : a \neq 2\}$ and $B = \{b \in \mathbf{R} : b \neq 1\}$. Define $f : A \to B$ by $f(a) = \frac{a}{a-2} \forall a \in A$. To show that the given function is one–one, let $a, b \in A$ such that $f(a) = f(b)$. Then we will show that $a = b$. If $f(a) = f(b)$, then we have

$$\frac{a}{a-2} = \frac{b}{b-2},$$

which implies that $ab - 2a = ab - 2b$, and hence $a = b$. So, the given function is one–one. Also, we have already shown that $R(f) = \mathbf{R} - \{1\} = B$. So this given function is one–one and onto.

Remark. If A and B are two **finite** sets, then $f : A \to B$ is one–one and onto if they have the same number of elements. If number of elements in A is less than the number of elements in B, then f can't be onto. If number of elements in B is less than the number of elements in A, then f can't be one–one. A one–one and onto function is also called a bijective function.

1.5 Composition of Functions

Let $f : A \to B$ and $g : B \to C$ be two functions and if $R(f) \subseteq$ Domain $(g) = B$, then their composition $g \circ f$ is a function from A to C and is defined as

$$(g \circ f)(a) = g(f(a)) \forall a \in A.$$

Example 1. Let $A = \{1, 2\}, B = \{3, 4, 5\}$, and $C = \{6, 7\}$. Define $f : A \to B$ by

$$f(1) = 3 \quad \text{and} \quad f(2) = 4,$$

$$\text{and} \quad g : B \to C \text{ by}$$

$$g(3) = 6, \quad g(4) = 7, \quad \text{and} \quad g(5) = 7.$$

Then the composite function $g \circ f : A \to C$ is given by

$$(g \circ f)(1) = g(f(1)) = g(3) = 6,$$

$$(g \circ f)(2) = g(f(2)) = g(4) = 7.$$

Example 2. Define $f : R \to R$ by $f(x) = x^2 - 1$ and $g : R \to R$ by $g(x) = 3x$, then the composite function $g \circ f : R \to R$ is given by

$$(g \circ f)(x) = g(f(x)) = g(x^2 - 1) = 3(x^2 - 1).$$

Inverse function. Inverse of a function does not always exist. If a function is one–one and onto, then its inverse exists. Let $f : A \to B$ be a bijective function. Then its inverse function $g : B \to A$ is defined as follows:

$$f^{-1}(B) = g(B) = \{a \in A : f(a) \in B\}.$$

Example. Let $A = \{1, 2\}$ and $B = \{3, 4\}$. Define $f : A \to B$ by $f(1) = 3$ and $f(2) = 4$.

Then the inverse function $f^{-1} : B \to A$ is given by $f^{-1}(3) = 1$ and $f^{-1}(4) = 2$.

1.6   Geometrical Interpretations

- If we draw a vertical line at $x = a$, where a is any point of the domain, then it will intersect the graph of f at most at one point. Then the graph represents a function.
- If we draw a horizontal line at $y = b$, where b is any point of codomain, then it will intersect the graph of f at most at one point. Then the function is one–one.
- If we draw a horizontal line at $y = b$, where b is any point of codomain, then it will intersect the graph of f at least at one point. Then the function is onto.

Chapter 2

Real Number System

In this chapter, we show how the set of real numbers is different from other sets (such as the sets of natural numbers, integers, rational numbers, and irrational numbers). There are three basic properties which make the set of real numbers different from other sets. These three properties are field axioms, ordered axioms, and completeness axioms. We show that the set of real numbers is the only set which is a complete ordered field.

First, we discuss field axioms which are based on two operations, addition $(+)$ and multiplication $(\cdot)$. Then we give the order properties, by using which we can solve inequalities in real analysis. We define absolute value by the use of the order property. In continuation, we give the definitions of bounded sets, supremum, and infimum. Finally, we discuss the most important property of the real number system known as the completeness axioms. Using the completeness axiom, we prove the Archimedean property.

2.1 Binary Operation

A map from $X \times X \to X$ which associates each ordered pair (x, y) in $X \times X$ to a unique point of X is called binary operation. In closure law, we operate two elements of a set and get some element of that set, which means closure law is also a binary operation. So if we write binary operation instead of operation, then it is not required to prove the closure property.

There are three basic axioms of a set.

1. Field axioms.
2. Order axioms.
3. Completeness axioms.

2.2   Field Axioms

Let F be a set with two binary operations: addition $(+)$ and multiplication $(\cdot)$. Then it satisfies the following properties.

a. Additive properties.

(a1) Associative law. $a + (b + c) = (a + b) + c$, $\forall a, b, c \in F$.
(a2) Existence of identity. There exist some $0 \in F$ such that $a + 0 = a = 0 + a$, $\forall a \in F$.
(a3) Existence of inverse. For each $a \in F$, there exists $-a \in F$ such that $a + (-a) = 0 = (-a) + a$, $\forall a \in F$.
(a4) Commutative property. $a + b = b + a$. $\forall a, b \in F$.

b. Multiplicative properties.

(b1) Associative law. $a \cdot (b \cdot c) = (a \cdot b) \cdot c$, $\forall a, b, c \in F$.
(b2) Existence of identity. There exists some $1 \in F$ such that $a \cdot 1 = a = 1 \cdot a$, $\forall a \in F$.
(b3) Existence of inverse. For each $a \in F$, there exists some $a^{-1} \in F$ such that $a \cdot a^{-1} = 1 = a^{-1} \cdot a$, $\forall a \in F$.
(b4) Commutative property. $a \cdot b = b \cdot a$, $\forall a, b \in F$.

c. Distributive properties.

(c1) $a \cdot (b + c) = a \cdot b + a \cdot c$, $\forall a, b, c \in F$.
(c2) $(a + b) \cdot c = a \cdot c + b \cdot c$, $\forall a, b, c \in F$.

Example. $\mathbf{R}$ and $\mathbf{Q}$ form a field. The set $\mathbf{N}$ of natural numbers does not form a field because the additive identity 0 is not a natural number. The set $\mathbf{Z}$ also does not form a field because multiplicative inverse a^{-1} does not exist for each integer.

   Order axioms refer to the notation of inequalities and positivity of a set. To solve the inequalities in a set, the order axioms play a very important role.

2.3 Order Axioms

Let X be a set with a relation $>$ and a, b, c be any elements of X then it satisfies the following properties:

(o1) Trichotomy property. If $a \neq b$, then either $a > b$ or $b > a$.
(o2) Transitive property. If $a > b$ and $b > c$, then $a > c$.
(o3) If $a > b$, then $a + c > b + c$.
(o4) If $a > b$ and $c > 0$, then $ac > bc$.

Example. The sets **N, Z, Q,** and **R** with the relation greater than ($>$) have ordered property. So the sets **Q** and **R** form **an** ordered field.

Theorem 1. *If $a \in R$, with $0 \leq a < \varepsilon$, for each $\varepsilon > 0$, then $a = 0$.*

Proof. We will prove it by contradiction. If possible, we let $a \neq 0$ and choose $\varepsilon = \frac{a}{3} > 0$. Then we have $0 < \frac{a}{3} < a$ which contradicts the fact that $a < \varepsilon$, for each $\varepsilon > 0$. So our assumption is wrong and, therefore, $a = 0$.

Theorem 2. *If $ab > 0$, then either*

a. $a < 0$ *and* $b < 0$ *or*
b. $a > 0$ *and* $b > 0$.

Proof. If $ab > 0$, then $a \neq 0$ and $b \neq 0$. Thus, by the Trichotomy property for a, either $a > 0$ or $a < 0$. Let $a < 0$. Then we will prove that $b < 0$. If $a < 0$, then $\frac{1}{a} < 0$, so that $b = \left(\frac{1}{a}\right)(ab) < 0$.
   Similarly, if $a > 0$, then $\frac{1}{a} > 0$, so that $b = \left(\frac{1}{a}\right)(ab) > 0$.

Example 1. Find a set $A = \{x \in \mathbf{R} : \quad x^2 - 5x > -6\}$.
   Let $x \in A \iff x^2 - 5x > -6 \iff x^2 - 5x + 6 > 0 \iff (x - 2)$ $(x - 3) > 0$.
   The following two cases arise:

i. $(x - 2) < 0$ and $(x - 3) < 0$, which imply that $x < 2$ and $x < 3$. We conclude that both inequalities are satisfied when $x < 2$.
ii. $(x - 2) > 0$ and $(x - 3) > 0$, which imply that $x > 2$ and $x > 3$. We conclude that both inequalities are satisfied when $x > 3$.

Combining the above two possibilities, the set A is given as follows:

$$A = \{x \in \mathbf{R}: \quad x < 2\} \cup \{x \in \mathbf{R}: \quad x > 3\}.$$

Example 2. Find a set $A = \{x \in \mathbf{R}: \quad \frac{3x+1}{x+3} < 2\}$.

Let $x \in A \iff \frac{3x+1}{x+3} < 2 \iff \frac{x-5}{x+3} < 0$.

The following two cases arise:

i. $(x - 5) < 0$ and $(x + 3) > 0$, which imply that $x < 5$ and $x > -3$. We conclude that both inequalities are satisfied when $-3 < x < 5$.
ii. $(x - 5) > 0$ and $(x + 3) < 0$, which imply that $x > 5$ and $x < -3$. We conclude that both inequalities can never be satisfied simultaneously.

So, only the first possibility is possible and the set A is given as follows:

$$A = \{x \in \mathbf{R}: -3 < x < 5\}.$$

Example 3. Find a set $A = \{x \in \mathbf{R}: \quad \frac{x+1}{x+3} < 2\}$.

Let $x \in A \iff \frac{x+1}{x+3} < 2 \iff \frac{-(x+5)}{x+3} < 0$.
The following two cases arise:

i. $(x+5) < 0$ and $(x+3) < 0$, which imply that $x < -5$ and $x < -3$. We conclude that both inequalities are satisfied when $x < -5$.
ii. $(x+5) > 0$ and $(x+3) > 0$, which imply that $x > -5$ and $x > -3$. We conclude that both inequalities are satisfied when $x > -3$.

Combining the above two possibilities, the set A is given as follows:

$$A = \{x \in \mathbf{R}: \quad x < -5\} \cup \{x \in \mathbf{R}: \quad x > -3\}.$$

Absolute value. The absolute value of a real number is the distance of that number from the origin and denoted as $|\cdot|$. It is given as follows:

$$|a| = \begin{cases} a & \text{if } a > 0, \\ -a & \text{if } a < 0, \\ 0 & \text{if } a = 0. \end{cases}$$

Properties of absolute value. Some concluding remarks about absolute value of real numbers are as follows:

   i. $|a| \geq 0$ and $|a| = 0$ iff $a = 0 \; \forall \, a \in \mathbf{R}$.
  ii. $a \leq |a|$ and $|a| = |-a| \; \forall \, a \in \mathbf{R}$.
 iii. $|a \cdot b| = |a| \cdot |b| \; \forall \, a, b \in \mathbf{R}$.
 iv. $|a|^2 = a^2 \; \forall \, a \in \mathbf{R}$.
  v. If $b \geq 0$, then $|a| \leq b$ iff $-b \leq a \leq b$.

     $|a| \leq b \Longleftrightarrow$ using (ii) we can write, $a \leq b$ and $-a \leq b \Longleftrightarrow -b \leq a \leq b$.

 vi. **Triangle inequality.** For all $a, b \in \mathbf{R}$, $|a + b| \leq |a| + |b|$.
     Using property (iv), we can write

$$|a + b|^2 = (a + b)^2 = a^2 + b^2 + 2ab,$$
$$\leq a^2 + b^2 + 2|ab|, \text{ using property (ii)},$$
$$= |a|^2 + |b|^2 + 2|ab|, \text{ using property (iv)}$$
$$= (|a| + |b|)^2.$$

Since both $|a + b|$ and $|a| + |b|$ are positive numbers, we can conclude that

$$|a + b| \leq |a| + |b|.$$

 vii. For all $a, b \in \mathbf{R}$, $|a - b| \leq |a| + |b|$.
     Using the triangle inequality for $a, -b \in \mathbf{R}$, we get

$$|a - b| \leq |a| + |-b|, \quad \text{since } |-b| = |b|.$$

     We thus conclude that $|a - b| \leq |a| + |b|$.
viii. For all $a, b \in \mathbf{R}$, $||a| - |b|| \leq |a - b|$.
     We can write $a = (a - b) + b$ and use the triangle inequality on both sides to get

$$|a| = |(a - b) + b| \leq |a - b| + |b|$$
$$\Rightarrow |a| - |b| \leq |a - b|.$$

Similarly, we can write $b = (b - a) + a$ and use the triangle inequality on both sides to get

$$|b| = |(b - a) + a| \leq |b - a| + |a|,$$
$$\Rightarrow |b| - |a| \leq |a - b|,$$
$$\Rightarrow -(|a| - |b|) \leq |a - b|,$$
$$\Rightarrow -(|a - b|) \leq |a| - |b|.$$

Since $|a - b| > 0$, combining two inequalities and using property (v), we conclude that

$$| \, |a| - |b| \, | \leq |a - b|.$$

2.4 Intervals

Open interval. Let $a, b \in \mathbf{R}$. Then an open interval is a set containing all the points between a and b. It is denoted by (a, b) and given as follows:

$$(a, b) = \{x \in \mathbf{R} : \quad a < x < b\}.$$

It is also called a segment.

Closed interval. Let $a, b \in \mathbf{R}$. Then a closed interval is a set containing all the points between a and b, and also includes the end points a and b. It is denoted by $[a, b]$ and given as follows:

$$[a, b] = \{x \in \mathbf{R} : \quad a \leq x \leq b\}.$$

Semi-open or semi-closed interval. Let $a, b \in \mathbf{R}$. Then a semi-open or semi-closed interval is a set containing all the points between a and b, and also includes the one end-point a or b. It is denoted by $(a, b]$ or $[a, b)$ and given as follows:

$$(a, b] = \{x \in \mathbf{R} : \quad a < x \leq b\},$$
$$[a, b) = \{x \in \mathbf{R} : \quad a \leq x < b\}.$$

Semi-infinite interval. Semi-infinite interval means one end point of the interval is infinity. Let $a \in \mathbf{R}$. Then a semi-infinite interval is a set containing all the points between a and ∞ or $-\infty$ and a. If it is semi-infinite open interval, then it does not include the end point a and if it is semi-infinite closed interval, then it includes the end point a. It is denoted by (a, ∞), $[a, \infty)$, $(-\infty, a)$ and $(-\infty, a]$ given as follows:

$$(a, \infty) = \{x \in \mathbf{R} : a < x\},$$
$$[a, \infty) = \{x \in \mathbf{R} : a \leq x\},$$
$$(-\infty, a) = \{x \in \mathbf{R} : x < a\},$$
$$(-\infty, a] = \{x \in \mathbf{R} : x \leq a\}.$$

Infinite interval. An interval whose both end points are infinity is the set of real numbers.

$$(-\infty, \infty) = \mathbf{R}.$$

2.5 Bounded Set

Upper bound of a set. An element $K \in \mathbf{R}$ is known as upper bound for a subset $A \subset \mathbf{R}$ if all the elements of A are less than or equal to K, that is,

$$a \leq K \quad \forall a \in A.$$

If a set A has an upper bound, then it is called a **bounded above** set.

Lower bound of a set. An element $k \in \mathbf{R}$ is known as lower bound for a subset $A \subset \mathbf{R}$ if all the elements of A are greater than or equal to k, that is,

$$k \leq a \quad \forall a \in A.$$

If a set A has a lower bound, then it is called a **bounded below** set.

Bounded set. A set A is said to be bounded if it is bounded below as well as bounded above, that is, there exist two real numbers k and K such that

$$k \leq a \leq K, \quad \forall a \in A.$$

We can also say that a set A is bounded if there exists a real number K such that

$$|a| \leq K, \quad \forall\, a \in A.$$

Example 1. The set of natural numbers $\mathbf{N} = \{1, 2, 3, \ldots\}$ is bounded below, but not bounded above.

Example 2. The set of negative integers $\mathbf{Z}^- = \{-1, -2, -3, \ldots\}$ is bounded above, but not bounded below.

Example 3. The set of integers $\mathbf{Z} = \{0, \pm 1, \pm 2, \pm 3, \ldots\}$ is neither bounded below nor bounded above.

Example 4. Set $A = \{\frac{1}{n} : n \in \mathbf{N}\} = \{1, \frac{1}{2}, \frac{1}{3}, \ldots\}$ is bounded.

2.6 Supremum and Infimum

Supremum. If a set A is bounded above, then we take the collection of all upper bounds of that set and the smallest upper bound is known as the supremum of the set. It is denoted by **Sup.** A. Let $K \in \mathbf{R}$ be supremum of a set A. Then

i. Since K is also an upper bound of the set, so $a \leq K, \quad \forall\, a \in A$.
ii. Let M be another upper bound of the set. Then $K \leq M$.

Infimum. If a set A is bounded below, then we take the collection of all lower bounds of that set and the greatest lower bound is known as the infimum of the set. It is denoted by **Inf.** A. Let $k \in \mathbf{R}$ be infimum of a set A then

i. Since k is also a lower bound of the set, $k \leq a, \quad \forall\, a \in A$.
ii. Let m be another lower bound of the set. Then $m \leq k$.

Remark. The supremum and infimum of a set may or may not exist. If a set is bounded, then its supremum and infimum will exist. It is also not required that the supremum and infimum of a set are members of that set. It may be from the set or may not be from the set.

Example 1. Set $A = \{-1, -\frac{1}{2}, -\frac{1}{3}, -\frac{1}{4}, \ldots\}$ is bounded.
Sup. $A = 0$ (does not belong to the set) **Inf.** $A = -1$ (belongs to the set).

Example 2. Set $A = \{\frac{1}{n} : n \in \mathbf{N}\}$ is bounded.
Sup. $A = 1$ (belongs to the set) **Inf.** $A = 0$ (does not belong to the set).

Example 3. Set $A = \{\frac{(-1)^n}{n} : n \in \mathbf{N}\}$ is bounded.
Sup. $A = \frac{1}{2}$ and **Inf.** $A = -1$ (both belong to the set).

Example 4. Set $A = \{\pm n^2 : n \in \mathbf{N}\}$. Both **Sup.** and **Inf.** of the set do not exist.

Theorem 1. *If the supremum of a non-empty subset $A \subset \mathbf{R}$ exists, then it is unique.*

Proof. Let, if possible, K_1 and K_2 be two supremums of a set A. So K_1 and K_2 will also be the upper bounds of A.

Since K_1 is the supremum of A and K_2 is an upper bound of A, by using the definition of a supremum (least upper bound), we can write

$$K_1 \leq K_2. \tag{2.1}$$

Similarly, since K_2 is the supremum of A and K_1 is an upper bound of A, again by using the definition of a supremum (least upper bound), we can write

$$K_2 \leq K_1. \tag{2.2}$$

From Eqs. (2.1) and (2.2), we get

$$K_1 = K_2.$$

So, the supremum is unique.

Theorem 2. *If the infimum of a non-empty subset $A \subset \mathbf{R}$ exists, then it is unique.*

Proof. Let, if possible, k_1 and k_2 be two infimums of A. Then k_1 and k_2 will also be the lower bounds of A.

Since k_1 is an infimum of A and since k_2 is a lower bound of A, by using the definition of an infimum (greatest lower bound), we can write

$$k_2 \leq k_1. \tag{2.3}$$

Similarly, since k_2 is the infimum of A and since k_1 is a lower bound of A, by using the definition of an infimum (greatest lower bound), we can write

$$k_1 \leq k_2. \tag{2.4}$$

From Eqs. (2.3) and (2.4), we get

$$k_1 = k_2.$$

So, the infimum is unique.

Theorem 3. *An upper bound K of a non-empty subset $A \subset \mathbf{R}$ is a supremum of A if and only if, for each $\varepsilon > 0$, there exists a real number $a \in A$ such that $K - \varepsilon < a$.*

Proof. Let K be an upper bound of a non-empty subset $A \subset \mathbf{R}$, and also satisfy the given condition $K - \varepsilon < a$. Then we will show that K is a supremum of A.

Since K is an upper bound, we only need to show that K is a least upper bound. For this, we will show that any number smaller than K cannot be an upper bound of A. Let, if possible, $k < K$ be an upper bound of A. Then, taking $\varepsilon = K - k > 0$ and using the given condition and the value of ε, we get

$$a > K - \varepsilon = K - (K - k) = k$$

$$\Rightarrow k < a.$$

So, k cannot be an upper bound of A.

Therefore, K is a least upper bound, that is, K is supremum of A.

Conversely, let K be supremum of A. If possible, there exists an $\varepsilon > 0$, such that $K - \varepsilon > a \quad \forall a \in A$. So $K - \varepsilon$ will be an upper bound of the set A, since $K - \varepsilon < K$.

But K is supremum of A, therefore, any number smaller than K cannot be an upper bound of A. So, our assumption that there exists an $\varepsilon > 0$ such that $K - \varepsilon > a \quad \forall a \in A$ is wrong.

Hence, for each $\varepsilon > 0$, there exists a real number $a \in A$ such that $K - \varepsilon < a$.

Theorem 4. *A lower bound k of a non-empty subset $A \subset \mathbf{R}$ is an infimum of A if and only if, for each $\varepsilon > 0$, there exists a real number $a \in A$ such that $a < k + \varepsilon$.*

Proof. Let k be a lower bound of a non-empty subset $A \subset \mathbf{R}$, which also satisfies the given condition $a < k + \varepsilon$. Then we will show that k is an infimum of A.

Since k is a lower bound, we only need to show that k is a greatest lower bound. For this, we will show that any number greater than k cannot be a lower bound of A. If possible, let $k < K$ be a lower bound of A. Then, taking $\varepsilon = K - k > 0$ and using the given condition and the value of our chosen ε, we get

$$a < k + \varepsilon = k + (K - k) = K$$
$$\Rightarrow a < K.$$

So, K cannot be a lower bound of A.

Therefore, k is a greatest lower bound, that is, k is the infimum of A.

Conversely, let k be the infimum of A. If possible. There exists an $\varepsilon > 0$ such that $k + \varepsilon < a \quad \forall a \in A$. So, $k + \varepsilon$ will be a lower bound of A, since $k < k + \varepsilon$.

But, k is infimum of A, therefore any number greater than k cannot be a lower bound of A. So, our assumption that there exists a $\varepsilon >$, such that $k + \varepsilon < a \quad \forall a \in A$, is wrong.

Hence, for each $\varepsilon > 0$, there exists a real number $a \in A$, such that $a < k + \varepsilon$.

Theorem 5. *Let A and B be two non-empty subsets of $\mathbf{R}$ that satisfy the following property:*

$$a \leq b \quad \forall a \in A \text{ and} \forall b \in B. \text{ Then } \textbf{\textit{Sup.}} \, A \leq \textbf{\textit{Inf.}} \, B.$$

Proof. Let b be an arbitrary element of B. Then, from the given condition:

$$a \leq b \quad \forall a \in A,$$

b is an upper bound of A. By the definition of upper bound and supremum, we can write

$$\textbf{Sup.} \, A \leq b.$$

Since b is an arbitrary element, the given inequality is true for all elements of B, that is,

$$\textbf{Sup.}\, A \leq b, \quad \forall\, b \in B,$$

So, from the above inequality, it is clear that **Sup.** A is a lower bound for the set B. Since infimum is the greatest lower bound, we can write

$$\textbf{Sup.}\, A \leq \textbf{Inf.}\, B.$$

As discussed above, we have two sets $\mathbf{Q}$ and $\mathbf{R}$, each of which is an ordered field. But $\sqrt{2}$ cannot be represented by a rational number. So, there is some extra property on the set of $\mathbf{R}$ in comparison with the set of $\mathbf{Q}$. This extra property of the set of $\mathbf{R}$ is known as completeness axiom.

2.7 Completeness Axiom

A set A is said to be complete if every non-empty bounded above subset of A has supremum in A. The set of real numbers is complete because its every non-empty bounded above subset has supremum in $\mathbf{R}$. But the set $\mathbf{Q}$ is not complete.

Theorem 1. *The set of rational numbers is not complete.*

Proof. To show that the set of rational numbers is not complete, we will show that there is a non-empty bounded above subset of $\mathbf{Q}$ which does not have supremum in $\mathbf{Q}$.

Consider a set $A = \{r \in \mathbf{Q}\colon r \geq 0 \text{ and } r^2 < 2\}$. Then, clearly, $A \subset \mathbf{Q}$, $1 \in A$ and A is bounded above by 2. So, A is a non-empty bounded above subset of $\mathbf{Q}$.

Now, we will show that A does not have supremum in $\mathbf{Q}$. Let if possible $l \in \mathbf{Q}$ be a supremum of A. Then by trichotomy property of ordered axiom, the three following possibilities arise:

(1) $l^2 < 2$ (2) $l^2 > 2$ (3) $l^2 = 2$.

Case 1. For $l^2 < 2$, we consider a rational number $x = \frac{4+3l}{3+2l}$. Then

$$2 - x^2 = 2 - \left(\frac{4+3l}{3+2l}\right)^2 = \frac{2-l^2}{9+12l+4l^2}. \tag{2.5}$$

Since l is the supremum of a positive set A, l is positive. Therefore, $9 + 12l + 4l^2 > 0$ and from our consideration $2 - l^2 > 0$.

Using these two inequalities in Eq. (2.5), we get

$$2 - x^2 > 0 \Rightarrow x^2 < 2,$$

$$\Rightarrow x \in A.$$

Also, we have

$$l - x = l - \frac{4 + 3l}{3 + 2l} = \frac{2(l^2 - 2)}{2l + 3}. \tag{2.6}$$

Again, since l is positive, therefore, $2l + 3 > 0$ and from our consideration $l^2 - 2 < 0$.

Using these two inequalities in Eq. (2.6), we get

$$l - x < 0 \Rightarrow l < x.$$

This contradicts the fact that l is the supremum of the set A because we have shown that $x \in A$ and $l < x$. So, $l^2 < 2$ is not possible.

Case 2. For $l^2 > 2$, we consider a rational number $x = \frac{4+3l}{3+2l}$. Then

$$2 - x^2 = 2 - \left(\frac{4 + 3l}{3 + 2l}\right)^2 = \frac{2 - l^2}{9 + 12l + 4l^2}. \tag{2.7}$$

Since l is positive, therefore, $9 + 12l + 4l^2 > 0$ and from our consideration $2 - l^2 < 0$.

Using these two inequalities in Eq. (2.7), we get

$$2 - x^2 < 0 \Rightarrow x^2 > 2$$

$\Rightarrow x$ is an upper bound of A.

Also, we have

$$l - x = l - \frac{4 + 3l}{3 + 2l} = \frac{2(l^2 - 2)}{2l + 3}. \tag{2.8}$$

Again, since l is positive, therefore, $2l + 3 > 0$ and from our consideration $l^2 - 2 > 0$.

Using these two inequalities in Eq. (2.8), we get

$$l - x > 0 \Rightarrow l > x.$$

This contradicts the fact that l is the supremum of the set A because we have shown that x is an upper bound of A and $l > x$. So, $l^2 > 2$ is not possible.

Case 3. If $l^2 = 2$, but there is no rational number whose square is 2, then $l^2 = 2$ is not possible.

From the above discussion, it is clear that all three cases are not possible. Therefore, $l \in \mathbf{Q}$ cannot be a supremum of A, that is, we have a non-empty bounded above subset of rational numbers which does not have supremum in $\mathbf{Q}$.

Therefore, the set of rational numbers is not complete.

Remark. The only complete ordered field is the set of real numbers. So, we can assume that any member of a complete ordered field is a real number.

2.8   Archimedean Property

If a and b are two positive real numbers, then there exists a natural number n such that

$$na > b.$$

Proof.   We will prove the Archimedean property by contradiction. Let a and b be two positive real numbers and, if possible, $na \leq b$, $\forall n \in \mathbf{N}$. Then b is an upper bound for the set,

$$A = \{na : \quad n \in \mathbf{N}\}.$$

A is a non-empty subset of real numbers which is bounded above by b. So, by the completeness axiom of real numbers, it has supremum in $\mathbf{R}$. Let K be the supremum of A.

Further, since $n + 1 \in \mathbf{N}$ and K is the supremum of A, so we can write

$$(n + 1)a \leq K$$

$$\Rightarrow na \leq K - a, \quad \forall n \in \mathbf{N}.$$

Thus, $K - a$ is an upper bound of A, which contradicts the fact that K is the supremum (least upper bound) of A because $K - a < K$. So, our assumption that $na \leq b$, $\forall\, n \in \mathbf{N}$, is false, that is, there exists a natural number n such that $na > b$.

Corollary 1. *The set of natural numbers is not bounded, that is, for a positive real number a, there exists a natural number n such that $n > a$.*

Proof. Let 1 and a be two positive real numbers. Then, by the Archimedean property, there exists a natural number n such that $n \cdot 1 > a$, that is, $n > a$.

Corollary 2. *For any $\varepsilon > 0$, there exists a natural number n such that $\frac{1}{n} < \varepsilon$.*

Proof. Since $\varepsilon > 0$, let $a = \varepsilon$ and $b = 1$ be two positive real numbers. Then, by the Archimedean property, there exists a natural number n such that $n\varepsilon > 1$, that is, $\varepsilon > \frac{1}{n}$.

2.9 Applications of the Archimedean Property

Theorem 1 (Density Theorem). *Between any two distinct real numbers a and b , there exists a rational number and hence infinitely many rational numbers.*

Proof. Let a and b be two distinct real numbers. Then, by the Trichotomy property of ordered axioms, either $a < b$ or $b < a$. Let $a < b$. Then we will prove that there exists a rational number r such that $a < r < b$.

Now, from our assumption that $b - a > 0$, $b - a$ and 1 are two positive real numbers. By the Archimedean property, there exists a natural number n such that

$$n(b - a) > 1,$$
$$\Rightarrow nb - na > 1. \tag{2.9}$$

From Eq. (2.9), it is clear that nb and na are two real numbers whose difference is greater than 1. Therefore, there always exists an integer between them. Let $m \in \mathbf{Z}$ be an integer which exists between nb and na. Then we can write

$$na < m < nb$$
$$\Rightarrow a < \frac{m}{n} < b.$$

Here m is an integer and n is a natural number. Therefore, we have

$$r = \frac{m}{n} \in \mathbf{Q}.$$
$$\Rightarrow a < r < b.$$

Now, we can apply the above theorem for two real numbers a and r, and we get a rational number r_1 lying between them, that is, $a < r_1 < r$.

Similarly, for r and b, we get a rational number r_2 lying between them, that is, $r < r_2 < b$.

So, $a < r_1 < r < r_2 < b$.

Continuing this process, we get infinite number of rational numbers between any two real numbers. Similarly, we can do for $b < a$.

So, the set of rational numbers is dense in the set of real numbers.

Theorem 2. *Between any two distinct real numbers a and b, there exists an irrational number and hence infinitely many irrational numbers.*

Proof. Let a and b be two distinct real numbers. Then, by the Trichotomy property of ordered axioms, either $a < b$ or $b < a$. Let $a < b$. Then we will prove that there exists an irrational number s such that $a < s < b$.

Now, from our assumption that $b - a > 0$, $b - a$ and $\sqrt{2}$ are two positive real numbers. By the Archimedean property, there exists a natural number n such that

$$n(b - a) > \sqrt{2}$$
$$\Rightarrow \frac{nb}{\sqrt{2}} - \frac{na}{\sqrt{2}} > 1. \qquad (2.10)$$

From Eq. (2.10), it is clear that $\frac{nb}{\sqrt{2}}$ and $\frac{na}{\sqrt{2}}$ are two real numbers whose difference is greater than 1. Therefore, there always exists an integer between them. Let $m \in \mathbf{Z}$ be the integer which exists between $\frac{nb}{\sqrt{2}}$ and $\frac{na}{\sqrt{2}}$. Then we can write

$$\frac{na}{\sqrt{2}} < m < \frac{nb}{\sqrt{2}},$$

$$\Rightarrow a < \frac{\sqrt{2}m}{n} < b.$$

m is an integer and n is a natural number, therefore, $s = \frac{\sqrt{2}m}{n}$ is an irrational number.

$$\Rightarrow a < s < b.$$

Now, we can apply the above theorem for two real numbers a and s, and we get an irrational number s_1 lying between them, that is, $a < s_1 < s$.

Similarly, for s and b, we get an irrational number s_2 lying between them, that is, $s < s_2 < b$.

So, $a < s_1 < s < s_2 < b$.

Continuing this process, we get infinite number of irrational numbers between any two real numbers. Similarly, we can do so for $b < a$.

So sets of irrational numbers are also dense in sets of real numbers.

2.10 Principle of Mathematical Induction

The principle of mathematical induction provides an important concept to prove a mathematical problem whose statement involves natural numbers. It is expressed as follows:

Let $P(n)$ be the given statement about n, $\forall\, n \in \mathbf{N}$. If

1. $P(1)$ is true.
2. $\forall\, l \in \mathbf{N},$ $P(l)$ is true, then $P(l+1)$ is true for all $l \in \mathbf{N}$.

Then $P(n)$ is true $\forall\, n \in \mathbf{N}$.

Example 1. The sum of first n natural numbers is given by

$$1 + 2 + 3 + \cdots + n = \frac{n(n+1)}{2}, \quad \forall\, n \in \mathbf{N}.$$

Proof. For $n = 1$,

$1 = \frac{1(1+1)}{2} = 1$, so it is true for $n = 1$. Let it be true for $l \in \mathbf{N}$. Then we have

$$1 + 2 + 3 + \cdots + l = \frac{l(l+1)}{2}.$$

Now, we will prove for $l + 1 \in \mathbf{N}$ that

$$1 + 2 + 3 + \cdots + l + (l+1) = \frac{(l+1)(l+2)}{2}.$$

L.H.S.

$$1 + 2 + 3 + \cdots + l + (l+1) = (1 + 2 + 3 + \cdots + l) + (l+1)$$

$$= \frac{l(l+1)}{2} + (l+1) = \frac{(l+1)(l+2)}{2}.$$

So, the given equality is true for all natural numbers.

Bernoulli's inequality. If $x > -1$, then show that $(1+x)^n \geq 1+nx$, $\forall\, n \in \mathbf{N}$.

Proof. For $n = 1$, $(1 + x)^1 \geq 1 + 1 \cdot x$, which is true.

Let $l \in \mathbf{N}$. Then we have

$$(1 + x)^l \geq 1 + lx. \tag{2.11}$$

Since $1 + x > 0$, by using the fact that $1 + x > 0$ and the following ordered axiom:

If $a > b$ and $c > 0$, then $ac > bc$)
in Eq. (2.11), we get

$$(1 + x)^l(1 + x) \geq (1 + lx)(1 + x).$$

Now, we will prove for $l + 1 \in \mathbf{N}$ that

$$(1 + x)^{l+1} = (1 + x)^l(1 + x).$$

From Eq. (2.11), we can write

$$(1 + x)^{l+1} = (1 + x)^l(1 + x) \geq (1 + lx)(1 + x) = 1 + lx + x + lx^2. \tag{2.12}$$

Since l is a natural number and $x^2 \geq 0$, so $lx^2 \geq 0$.

We can rewrite Eq. (2.12) as follows:

$$(1+x)^{l+1} \geq 1 + lx + x + lx^2 \geq 1 + (l+1)x,$$
$$\Rightarrow (1+x)^{l+1} \geq 1 + (l+1)x.$$

So, the given inequality is true for all natural numbers.

Exercises

1. Find a set $A = \{x \in \mathbf{R} : \quad x^2 + x > 2\}$.

   Ans. $A = \{x \in \mathbf{R} : x > 1\} \cup \{x \in \mathbf{R} : x < -2\}$.

2. Find a set $D = \left\{x \in \mathbf{R} : \quad \frac{2x+1}{x+2} < 1\right\}$.

   Ans. $D = \{x \in \mathbf{R} : -2 < x < 1\}$.

3. Show that for any two real numbers a and b,

$$\max\{a, b\} = \frac{1}{2}(a + b + |a - b|).$$

4. Find the supremum and infimum of the following sets:

   a. $A = \{-\frac{1}{n} : n \in \mathbf{N}\}$.

   b. $A = [1, 2) \cup [3, 7)$.

   c. $A = \left\{\left(1 - \frac{1}{n}\right) \sin \frac{n\pi}{2} : n \in \mathbf{N}\right\}$.

   d. $A = \{x \in \mathbf{I} : x^2 \leq 36\}$.

   Ans. a. Sup. $A = 0$, Inf. $A = -1$.   b. Sup. $A = 7$, Inf. $A = 1$.
   c. Sup. $A = 1$, Inf. $A = -1$.   d. Sup. $A = 6$, Inf. $A = -6$.

Chapter 3

Basics of Real Analysis

The present chapter is very important for the study of the subsequent chapters. In this chapter, first we study the neighborhoods. We then define the interior points using which the definition of open set is given. Some basic theorems of open sets of real numbers are described. Further, we give the definition of limit points, derived sets, adherent points, and closed sets. We deal briefly with the properties of closed sets. We also give the definition of dense and perfect sets. Finally, we discuss countable and uncountable sets and some examples of these sets.

3.1 Neighborhood of a Point

Let $a \in \mathbf{R}$. Then a set N is said to be a neighborhood of point a if there exists an open interval I containing a and contained in N. So the neighborhood is a set and considered for a point. It is denoted by "**nbd**" and is given as follows:

$$a \in I \subseteq N.$$

An open interval $(a - \varepsilon, a + \varepsilon)$, where $\varepsilon > 0$, is the **nbd** of a point a because

$$a \in I = (a - \varepsilon, a + \varepsilon) \subseteq N = (a - \varepsilon, a + \varepsilon).$$

The **nbd** of a point $a \in \mathbf{R}$ is a set given by $\{x \in \mathbf{R} : |x - a| < \varepsilon\}$.

Example 1. A finite set, which is non-empty, is not the nbd of any point in it.

Let A be a finite set and be the **nbd** of a point $a \in A$. Then, by definition of the **nbd**, there exists an open interval $(a - \varepsilon, a + \varepsilon)$, where $\varepsilon > 0$, such that

$$a \in (a - \varepsilon, a + \varepsilon) \subseteq A,$$

which is not possible, because a finite set A cannot contain an infinite set $(a - \varepsilon, a + \varepsilon)$. So, our assumption that a finite set is the **nbd** of its point is false. A non-empty finite set is not the **nbd** of its any point.

Example 2. An open interval (a, b) is the **nbd** of all its points, but a closed interval $[a, b]$ is the **nbd** of the set (a, b) That is, $[a, b]$ is not the **nbd** of its end points a and b.

Let $(a - \varepsilon, a + \varepsilon)$, where $\varepsilon > 0$, be an open interval containing a, and if $[a, b]$ is the **nbd** of a, then

$a \in (a - \varepsilon, a + \varepsilon) \subseteq [a, b]$, which is not true. Similarly, we can show it for the point b.

Example 3. The set $\mathbf{Q}$ of rational numbers is not the **nbd** of any of its points.

Let $\mathbf{Q}$ be the **nbd** of a point $r \in \mathbf{Q}$. Then, by the definition of the **nbd**, there exists an open interval $(r - \varepsilon, r + \varepsilon)$, where $\varepsilon > 0$, such that

$$r \in (r - \varepsilon, r + \varepsilon) \subseteq \mathbf{Q},$$

which is not possible, because the open interval $(r - \varepsilon, r + \varepsilon)$ contains an infinite number of irrational numbers (irrationals are dense in real). So, it cannot be a subset of $\mathbf{Q}$. Therefore, $\mathbf{Q}$ is not the **nbd** of any of its points.

Example 4. The set $\mathbf{R}$ of real numbers and an empty set are the **nbd** of all its points.

Properties of neighborhood.

1. Superset of the **nbd** of a point is also the **nbd** of that point.
2. Intersection of the two neighborhoods of a point is also the neighborhood of that point.
3. The union of the two neighborhoods of a point is also the neighborhood of that point.

3.2 Interior Points

Let $a \in A$. Then the point a is an interior point of A if there exists an open interval, which contains a and is contained in A, that is, $a \in (a - \varepsilon, a + \varepsilon) \subseteq A$, where $\varepsilon > 0$. The point outside the set cannot be an interior point.

Interior of a set. The interior of a set A is the collection of all interior points of A and it is denoted by int A.

Example 1. The set $\mathbf{N}$ of natural numbers does not have any interior point. Let $a \in \mathbf{N}$ be an interior point of $\mathbf{N}$. Then, by definition of the interior point, there exists an open interval $(a - \varepsilon, a + \varepsilon)$, where $\varepsilon > 0$, such that

$$a \in (a - \varepsilon, a + \varepsilon) \subseteq \mathbf{N},$$

which is not possible, because the interval $(a - \varepsilon, a + \varepsilon)$ contains points from the real line other than natural numbers. So, our assumption that the set $\mathbf{N}$ has an interior point is false.

Example 2. The sets $\mathbf{Z}$ and $\mathbf{Q}$ do not have any interior point. Therefore, int $\mathbf{Z} = \Phi$ and int $\mathbf{Q} = \Phi$.

Example 3. Let $A = (a, b)$ and $B = [a, b]$. Then int $A = (a, b)$ and int $B = (a, b)$. In $[a, b]$, the end points are not interior points.

3.3 Open Set

The set A is open if it is a **nbd** of all its points. That is, it contains all its interior points. If A is an open set, then, for all $a \in A$, there exists an open interval $(a - \varepsilon, a + \varepsilon)$, where $\varepsilon > 0$, such that

$$a \in (a - \varepsilon, a + \varepsilon) \subseteq A.$$

Therefore, for an open set A, int $A = A$.

Example 1. Since any point of the sets $\mathbf{N}$, $\mathbf{Z}$ and $\mathbf{Q}$ is not an interior point, these sets are not open.

Example 2. Since int $(a, b) = (a, b)$, therefore, (a, b) is an open set. But int $[a, b] = (a, b)$. Therefore, $[a, b]$ is not an open set.

Example 3. A non-empty finite set A is not open because int $A = \Phi$.

Theorem 1. *Every open interval is an open set, but the converse is not true.*

Proof. Let (a, b) be an open interval. In order to show that it is an open set, we show all its points are interior points. Let $k \in (a, b)$ be an arbitrary point.

Choose $\varepsilon = \min\{k - a, \; b - k\}$. Since $a < k < b$, so $k - a$ and $b - k$ are positive. Therefore, $\varepsilon > 0$. It is obvious from the construction of ε that

$$k \in (k - \varepsilon, \; k + \varepsilon) \subseteq (a, b).$$

This implies that k is an interior point. Since k is an arbitrary point, each point of (a, b) is an interior point.

Consequently, (a, b) is an open set.

To show that the converse is not true, we give an example of a set which is an open set but not an open interval.

Consider $A = (1, 2) \cup (3, 4)$. Since int $A = A$, A is an open set. But it is not an open interval, because it does not contain all points between 1 and 4.

Theorem 2. *Every open set is a union of open intervals.*

Proof. Let A be an open set. Then we will prove that A can be written as a union of open intervals. Let $a_\lambda \in A$ be an arbitrary point. Since A is open, there exists an open interval $(a_\lambda - \varepsilon, \; a_\lambda + \varepsilon)$, where $\varepsilon > 0$, such that

$$a_\lambda \in (a_\lambda - \varepsilon, \; a_\lambda + \varepsilon) \subseteq A, \quad \forall \, a_\lambda \in A.$$
$$\Rightarrow \bigcup_{\lambda \in \wedge} (a_\lambda - \varepsilon, \; a_\lambda + \varepsilon) \subseteq A,$$

where $\wedge$ is an index set. $\qquad(3.1)$

Since $a_\lambda \in (a_\lambda - \varepsilon, \; a_\lambda + \varepsilon) \; \forall \, a_\lambda \in A$ and since we can also write any set in the union of singleton sets, therefore, we have

$$A = \bigcup_{\lambda \in \wedge} \{a_\lambda\} \subseteq \bigcup_{\lambda \in \wedge} (a_\lambda - \varepsilon, \; a_\lambda + \varepsilon). \qquad(3.2)$$

From Eqs. (3.1) and (3.2), we can write

$$A = \bigcup_{\lambda \in \Lambda} (a_\lambda - \varepsilon, \, a_\lambda + \varepsilon).$$

So, every open set can be written as a union of open intervals.

Theorem 3. *The intersection of finite numbers of open sets is open.*

Proof. First, we will prove this theorem for the intersection of two open sets. Then it can be extended to finite numbers of open sets.

Let, A and B be two open sets. Then we will show that $A \cap B$ is an open set. Let $x \in A \cap B$

$$\Rightarrow \quad x \in A \quad \text{and} \quad x \in B.$$

Since A and B are open sets, by definition of open sets, there exist $\varepsilon_1 > 0$ and $\varepsilon_2 > 0$ such that

$$x \in (x - \varepsilon_1, \, x + \varepsilon_1) \subseteq A \quad \text{and} \quad x \in (x - \varepsilon_2, \, x + \varepsilon_2) \subseteq B.$$

Let $\varepsilon = \min\{\varepsilon_1, \, \varepsilon_2\}$, so $\varepsilon > 0$. Then

$$x \in (x - \varepsilon, \, x + \varepsilon) \subseteq A \quad \text{and} \quad x \in (x - \varepsilon, \, x + \varepsilon) \subseteq B.$$
$$\Rightarrow x \in (x - \varepsilon, \, x + \varepsilon) \subseteq A \cap B.$$

So, x is an interior point of $A \cap B$. Since x is an arbitrary point of $A \cap B$, therefore, each point of $A \cap B$ is an interior point, and hence $A \cap B$ is an open set. Similarly, we can prove the theorem for finite numbers of open sets.

Remark. The intersection of arbitrary numbers of open sets need not be open. Consider

$$A_n = \left(-\frac{1}{n}, \frac{1}{n} \right), \quad n \in \mathbf{N}.$$

Each A_n is an open set and $\bigcap_{n=1}^{\infty} A_n = \{0\}$. But $\{0\}$ is not an open set, because there does not exist any $\varepsilon > 0$ such that $(0 - \varepsilon, 0 + \varepsilon) \subseteq \{0\}$, since $(0 - \varepsilon, \, 0 + \varepsilon)$ is an infinite set and $\{0\}$ is a finite set.

Theorem 4. *The union of arbitrary numbers of open sets is open.*

Proof. Let $\{A_\lambda\}_{\lambda \in \Lambda}$ be an arbitrary collection of open sets. Then we will prove that $A = \cup_{\lambda \in \Lambda} A_\lambda$ is an open set.

Let $x \in A = \cup_{\lambda \in \Lambda} A_\lambda$. Then we will show that there exists an open interval containing x and contained in A.

$$x \in A = \bigcup_{\lambda \in \Lambda} A_\lambda \Rightarrow x \in A_\lambda \text{ for at least one } \lambda \in \Lambda.$$

Since A_λ is an open set, there exists an open interval $(x - \varepsilon, x + \varepsilon)$ such that

$$x \in (x - \varepsilon, x + \varepsilon) \subseteq A_\lambda,$$

$$\Rightarrow x \in (x - \varepsilon, x + \varepsilon) \subseteq A_\lambda \subseteq \bigcup_{\lambda \in \Lambda} A_\lambda$$

$$\Rightarrow x \in (x - \varepsilon, x + \varepsilon) \subseteq A = \bigcup_{\lambda \in \Lambda} A_\lambda,$$

Since x is an arbitrary point, each point of A is an interior point and, therefore, $A = \cup_{\lambda \in \Lambda} A_\lambda$ is an open set.

3.4 Limit Points

Let $A \subseteq \mathbf{R}$. Then a number $a \in \mathbf{R}$ is called a limit point of A if any neighborhood of a contains a point of A different from a. That is, for every $\varepsilon > 0$, we have

$$A \cap (a - \varepsilon, a + \varepsilon) - \{a\} \neq \Phi.$$

Derived set. The derived set of a set A is the collection of all limit points of A and it is denoted by der A or A'.

Example 1. The set $A = \left\{ \frac{1}{n} : n \in N \right\}$. Then 0 is the only limit point of the set A. If we take any neighborhood of 0, then it contains points of A different from 0, that is, $A \cap (0 - \varepsilon, 0 + \varepsilon) - \{0\} \neq \Phi$. Clearly, the intersection will contain the infinitely many points of A close to $0 - \varepsilon$. But a number of type $\frac{1}{n}$ cannot be its limit point, because, if we take $\varepsilon = \min\left\{ \frac{1}{n} - \frac{1}{n+1}, \frac{1}{n-1} - \frac{1}{n} \right\}$, then $\epsilon > 0$ and $A \cap \left(\frac{1}{n} - \varepsilon, \frac{1}{n} + \varepsilon \right) - \left\{ \frac{1}{n} \right\} = \Phi$. So, is the only limit point of the set A. Thus, der $A = \{0\}$.

Example 2. The set of integers $\mathbf{Z}$ has no limit point. Let $a \in \mathbf{Z}$, then its neighborhood $\left(a - \frac{1}{3}, a + \frac{1}{3}\right)$ has no point common from $\mathbf{Z}$ other than a. der $\mathbf{Z} = \Phi$. This is an example of a set which has no limit point.

Example 3. Consider the set $\mathbf{Q}$ of rational numbers. Then, if we take any neighborhood of $a \in \mathbf{R}$ it contains many points of $\mathbf{Q}$ different from a. Therefore, der $\mathbf{Q} = \mathbf{R}$. This is an example of a set which has infinite number of limit points.

Example 4. der $\mathbf{R} = \mathbf{R}$, der $\mathbf{N} = \Phi$, der $(a, b) = [a, b]$, and der $[a, b] = [a, b]$.

A finite set has no limit point. An infinite set may or may not have limit points. Now, we will state the Bolzano–Weierstrass theorem, which gives sufficient condition for a set to have a limit point.

Bolzano–Weierstrass theorem. *Every infinite bounded set has a limit point.*

Proof. Let S be any infinite bounded set and suppose that k and K are its infimum and supremum, respectively. Let A be a set of real numbers defined as follows:

$$A = \{x : x \text{ exceeds at most a finite number of members of } S\}.$$

The set A is non-empty for $k \in A$. Also K is an upper bound of A and any number greater than or equal to K cannot be a member of A. Thus, A is a non-empty bounded above subset of real numbers. Then, by the completeness property, its supremum exists. Let a be the supremum of A. We shall now show that a is the limit point of S. Consider any nbd $(a - \varepsilon, a + \varepsilon)$ of a, where $\varepsilon > 0$.

Since a is the supremum of A, $\exists$ at least one member, say b of A, such that $b > a - \varepsilon$. Since $b \in A$, it exceeds at most a finite number of members of S. Consequently, $b > a - \varepsilon$ can exceed at most a finite number of members of S.

Since a is the supremum of A, $a + \varepsilon$ cannot belong to A and consequently, $a + \varepsilon$ must exceed an infinite number of members of S.

Now, $a - \varepsilon$ exceeds at most a finite number of members of S and $a + \varepsilon$ exceeds infinitely many members of S.

$\Rightarrow$ $(a - \varepsilon, a + \varepsilon)$ contains an infinite number of member of S.

Therefore, a is a limit point of S.

Remark.

1. The boundedness condition is not superfluous in the Bolzano–Weierstrass theorem, because the set of integers is an infinite unbounded set which has no limit point.
2. A finite set has no limit point. So, this theorem is for infinite sets.
3. The Bolzano–Weierstrass theorem is sufficient, but not a necessary, condition for the existence of the limit points. Consider the set $A = \left\{ \frac{1}{2}, 4, \frac{1}{3}, 9, \frac{1}{4}, 16, \ldots \right\}$. Then A is a unbounded set and has as 0 its limit point.

Adherent points. Let $A \subseteq \mathbf{R}$. Then a number $a \in \mathbf{R}$ is called an adherent point of A if any neighborhood of a contains a point of A. That is, for every $\varepsilon > 0$, we have

$$A \cap (a - \varepsilon, a + \varepsilon) \neq \Phi.$$

So, from the definition, it is clear that any point of the set, that is, $a \in A$ is an adherent point because, if $a \in A$, then $A \cap (a - \varepsilon, a + \varepsilon) \neq \Phi$, so that it always contains at least one point a. By the definition of limit points, it is clear that each limit point is an adherent point.

Closure of a set. The closure of a set A is the collection of all adherent points of A and it is denoted by $\tilde{A}$.

$$\tilde{A} = A \cup A'.$$

3.5 Closed Sets

A set is said to be closed if it contains each of its limit points. We can also say that a set A is closed if $\tilde{A} = A$.

Thus, a set A is closed if and only if $A' \subseteq A$.

Example 1. Consider the set $A = \left\{ \frac{1}{n} : n \in N \right\}$. Then 0 is the only limit point of the set A, which is not a member of A. Therefore, A is not a closed set.

Example 2. The set $\mathbf{Z}$ of integers has no limit point. Since $\mathbf{Z}' = \Phi \subset \mathbf{Z}$, therefore, $\mathbf{Z}$ is a closed set.

Example 3. Consider the set $\mathbf{Q}$ of rational numbers. Since $\mathbf{Q}' = \mathbf{R} \not\subseteq \mathbf{Q}$, therefore, $\mathbf{Q}$ is not a closed set.

Example 4. Since $\mathbf{N}' = \mathbf{R} \subseteq \mathbf{R}$, $\mathbf{N}' = \Phi \subset \mathbf{N}$, and $[a,b]' = [a,b] \subseteq$ $[a,b]$, therefore, the sets $\mathbf{R}$, $\mathbf{N}$, and $[a,b]$ are closed sets.

Theorem 1. *A set is closed if and only if its complement is open.*

Proof. Let A be a closed set. Then we will show that its complement $A' = \mathbf{R} - A = B$ is an open set. To show that B is an open set, we need to show that each of its points is an interior point. Let $x \in B$ be an arbitrary point. Then $x \notin A$. Since A is a closed set, x cannot be its limit point. Therefore, there exists an open interval $(x - \varepsilon,\, x + \varepsilon)$ such that

$$A \cap (x - \varepsilon,\, x + \varepsilon) = \Phi,$$
$$\Rightarrow (x - \varepsilon,\, x + \varepsilon) \subseteq B,$$
$$\Rightarrow x \in (x - \varepsilon,\, x + \varepsilon) \subseteq B.$$

So x is an interior point of the set B. Since x is an arbitrary point, each point of the set B is an interior point. Hence, B is an open set.

Conversely, let $A' = \mathbf{R} - A = B$ be an open set. Then we will show that A is a closed set. To show that A is closed, we need to show that A has each of its limit points. That is, no point outside A is a limit point of A. Let, if possible, the point $x \notin A$ be a limit point of A. Since $x \notin A$ therefore, $x \in B$ and B is an open set. So there exists an open interval $(x - \varepsilon,\, x + \varepsilon)$ such that

$$x \in (x - \varepsilon,\, x + \varepsilon) \subseteq B$$
$$\Rightarrow A \cap (x - \varepsilon,\, x + \varepsilon) = \Phi,$$

which contradicts the fact that x is a limit point of A. So, a point outside A cannot be a limit point of A. Therefore, A is a closed set.

Theorem 2. *The intersection of an arbitrary collection of closed sets is a closed set.*

Proof. Let $\{A_\lambda\}_{\lambda \in \Lambda}$, be an arbitrary collection of closed sets. Then we will prove that $A = \bigcap_{\lambda \in \Lambda} A_\lambda$ is a closed set. To show that A is closed, we will show that its complement is open. That is, A' is open.

Using De Morgan's law, we can write

$$A' = \bigcup_{\lambda \in \wedge} A'_\lambda. \tag{3.3}$$

Since $\{A_\lambda\}_{\lambda \in \wedge}$ is an arbitrary collection of closed sets, A'_λ is an open set $\forall \lambda \in \wedge$. In Eq. (3.3), A' is the union of an arbitrary collection of open sets. Therefore, it is an open set. Hence, A is a closed set.

Theorem 3. *The union of a finite number of closed sets is a closed set.*

Proof. First, we will prove this theorem for the union of two closed sets. Then it can be extended to a finite number of closed sets.

Let, A and B be two closed sets. Then we will show that $A \cup B$ is a closed set. By De Morgan's law, we can write

$$(A \cup B)' = A' \cap B'. \tag{3.4}$$

Since A and B are closed sets, then A' and B' are open sets. Since the right-hand side in Eq. (3.4) is an intersection of two open sets, it is open. So $(A \cup B)'$ is an open set. Hence, $A \cup B$ is a closed set.

Remark. The union of arbitrary numbers of closed sets need not be closed. Consider

$$A_n = \left[\frac{1}{n}, 1 \right], \quad n \in \mathbf{N}.$$

Then each A_n is a closed set. But their union
$A = \cup_{n=1}^{\infty} A_n = (0, 1]$ is not a closed set, because is the limit point of the set A, but $0 \notin (0, 1]$.

3.6 Dense Sets

A set A is said to be dense in a set B if $\tilde{A} = B$. Let $A \subseteq \mathbf{R}$. Then A is said to be dense in $\mathbf{R}$ if $\tilde{A} = \mathbf{R}$ and it is said to be dense in itself if $\tilde{A} = A$.

Perfect sets. A set A is said to be perfect if it is closed as well as dense in itself.

Example 1. Let $A = [a, b]$. Then $\tilde{A} = A$. Therefore, $[a, b]$ is dense in itself. Since $[a, b]$ is closed and dense in itself, therefore, $[a, b]$ is a perfect set.

Example 2. Since $\tilde{\mathbf{Q}} = \mathbf{R}$, therefore, $\mathbf{Q}$ is dense in $\mathbf{R}$. $\mathbf{Q}$ is neither closed nor dense in itself, so $\mathbf{Q}$ is not a perfect set.

Example 3. Since $\tilde{\mathbf{R}} = \mathbf{R}$, and since $\mathbf{R}$ is also closed, $\mathbf{R}$ is a perfect set.

3.7 Countable and Uncountable Sets

Finite and infinite sets. A set A is finite if it is empty or it has a finite number n of elements. A set A has n elements if there exists a bijection from the set $\{1, 2, 3, \ldots, n\}$ onto A. A set which is not finite is known as an infinite set.

Countably infinite or denumerable sets. A set A is countably infinite or denumerable if there exists a bijection from the set of natural numbers onto A. Since the inverse image of a bijection map is again a bijection, a set A is countably infinite or denumerable if there exists a bijection from the set A onto the set of natural numbers.

Countable and uncountable sets. A set A is countable if it is either finite or countably infinite. A set A is uncountable if it is not countable.

Example 1. The set of integers $\mathbf{Z}$ is a denumerable set. Define a map $f : \mathbf{N} \to \mathbf{Z}$ by

$$f(n) = \begin{cases} 0, & \text{if } n = 1, \\ \text{negative integer}, & \text{if } n \text{ is odd}, \\ \text{positive integer}, & \text{if } n \text{ is even}, \end{cases}$$

then the map $f : \mathbf{N} \to \mathbf{Z}$ is a bijection. Therefore, the set $\mathbf{Z}$ of integers is a denumerable set.

Example 2. The set of even natural numbers $A = \{2n : n \in \mathbf{N}\}$ is a denumerable set. Define a map $f : \mathbf{N} \to A$ by

$$f(n) = 2n, \quad \forall\, n \in \mathbf{N},$$

then the map $f : \mathbf{N} \to A$ is a bijection. Therefore, the set A is a denumerable set.

Theorem. *Let A and B be two sets such that $A \subseteq B$.*

1. *If B is a countable set, then A is a countable set.*
2. *If A is an uncountable set, then B is an uncountable set.*

Sometimes it is not possible to define a bijection map explicitly. Then we "enumerate" the elements of the set. Enumeration means that we can find some arrangement for the elements of the set, that is, we can determine the next element of the set. If enumeration of the elements of a set is possible, then the set is countable.

Theorem 1. *The set $\mathbf{N} \times \mathbf{N}$ is countable.*

Proof. The set $\mathbf{N} \times \mathbf{N} = \{(a,b) : a, b \in \mathbf{N}\}$.

The enumeration for the elements of $\mathbf{N} \times \mathbf{N}$ is given as follows:

$$\underbrace{(1,1)}_{a+b=2}, \underbrace{(1,2),(2,1)}_{a+b=3}, \underbrace{(1,3),(2,2),(3,1)}_{a+b=4}, \underbrace{(1,4),(2,3),(3,2),(4,1)}_{a+b=5}, \ldots$$

These are arranged according to the increasing sum $a + b$ and increasing a. Since enumeration for the elements of $\mathbf{N} \times \mathbf{N}$ is possible, $\mathbf{N} \times \mathbf{N}$ is countable.

Theorem 2. *If A_n is countable for each $n \in \mathbf{N}$, then $A = \cup_{n=1}^{\infty} A_n$ is countable. That is, a countable union of countable sets is countable.*

Proof. In this proof, we use the process of enumeration. The elements of A_n are enumerated as follows:

$$A_1 = \{a_{11}, a_{12}, a_{13}, \ldots\},$$
$$A_2 = \{a_{21}, a_{22}, a_{23}, \ldots\},$$
$$A_3 = \{a_{31}, a_{32}, a_{33}, \ldots\},$$

$$\ldots \quad \ldots \quad \ldots$$

The set $A = \cup_{n=1}^{\infty} A_n = \{a_{ij} : i, j \in \mathbf{N}\}$ is enumerated as follows:

$$A = \left\{ \underbrace{a_{11}}_{i+j=2}, \underbrace{a_{12}, a_{21}}_{i+j=3}, \underbrace{a_{13}, a_{22}, a_{31}}_{i+j=4}, \underbrace{a_{14}, a_{23}, a_{32}, a_{41}}_{i+j=5}, \ldots \right\}$$

These are arranged according to the increasing sum $i + j$ and increasing i. Since enumeration for the elements of A is possible, A is countable.

Theorem 3. *The set $\mathbf{Q}$ of rational numbers is countable.*

Proof. The set $\mathbf{Q}$ of rational numbers can be written as the union of positive rational numbers, negative rational numbers, and the singleton $\{0\}$, that is,

$$\mathbf{Q} = \mathbf{Q}^+ \cup \mathbf{Q}^- \cup \{0\}.$$

The set of positive rational numbers, given by

$$\mathbf{Q}^+ = \left\{ \frac{p}{q} \ : \ q \neq 0 \text{ and } p, \ q \in \mathbf{Z}^+ \right\},$$

can be enumerated as follows:

$$\mathbf{Q}^+ = \left\{ \underbrace{\frac{1}{1}}_{p+q=2}, \ \underbrace{\frac{1}{2}, \frac{2}{1}}_{p+q=3}, \underbrace{\frac{1}{3}, \frac{3}{1}}_{p+q=4}, \underbrace{\frac{1}{4}, \frac{2}{3}, \frac{3}{2}, \frac{4}{1}}_{p+q=5}, \cdots \right\}.$$

These are arranged according to the increasing sum $p+q$ and increasing p (there is no common factor in p and q). Since enumeration for the elements of $\mathbf{Q}^+$ is possible, $\mathbf{Q}^+$ is countable. Similarly, we can show that the set $\mathbf{Q}^-$ of negative rational numbers is countable and $\{0\}$ is a finite set. Therefore, it is also countable.

Since $\mathbf{Q}$ is the union of three countable sets, it is countable.

Theorem 4. *The closed interval $[0, 1]$ is uncountable.*

Proof. We will prove it by contradiction. Let, if possible, $[0, 1]$ be countable. Then we can enumerate the elements of $[0, 1]$ as $\{a_1, a_2, a_3, a_4, a_5, \ldots, a_n, \ldots\}$. Since each $a_i \in [0, 1]$, the decimal representation for each a_i is given as follows:

$$a_1 = 0.x_{11}x_{12}x_{13} \ \cdots \ x_{1n} \cdots$$

$$a_2 = 0.x_{21}x_{22}x_{23} \ \cdots \ x_{2n} \cdots$$

$$a_3 = 0.x_{31}x_{32}x_{33} \ \cdots \ x_{3n} \cdots$$

$$\cdots \ \cdots \ \cdots \ \cdots \ \cdots$$

$$a_n = 0.x_{n1}x_{n2}x_{n3} \cdots x_{nn} \cdots$$

$$\cdots \ \cdots \ \cdots \ \cdots \ \cdots$$

where $x_{ij} \in \{0, 1, 2, \ldots, 9\}$. Consider a number

$$b = 0.y_1 y_2 y_3 \cdots y_n \cdots$$

where $y_i = \begin{cases} 1, & \text{if } x_{ii} = 2, \\ 2, & \text{if } x_{ii} \neq 2. \end{cases}$

Then b differs from a_1 at the first decimal place because $y_1 \neq x_{11}$ and b differs from a_2 at the second decimal place because $y_2 \neq x_{22}$. Similarly, b differs from each a_i at the ith decimal place because $y_i \neq x_{ii}$. So $b \in [0, 1]$ but it is different from each a_i which contradicts the fact that $\{a_1, a_2, a_3, a_4, a_5, \ldots, a_n, \ldots\}$ is enumeration for $[0, 1]$. So our assumption that $[0, 1]$ is countable is false. Therefore, $[0, 1]$ is uncountable.

Theorem 5. *The set $\mathbf{R}$ of real numbers is uncountable.*

Proof. Since $[0, 1] \subset \mathbf{R}$ and $[0, 1]$ is an uncountable set, the set $\mathbf{R}$ of real numbers is an uncountable set.

Theorem 6. *The set of irrational numbers is uncountable.*

Proof. Let A be the set of irrational numbers. Let, if possible, A be a countable set. We can write the set of real numbers as the union of rational and irrational numbers. That is,

$$\mathbf{R} = \mathbf{Q} \cup A.$$

Since $\mathbf{Q}$ and A both are countable, their union is also countable. That is, $\mathbf{R}$ is countable. This is a contradiction because $\mathbf{R}$ is uncountable. Therefore, our assumption that the set A is countable is false and hence the set A of irrational numbers is uncountable.

Exercises

1. Obtain the derived set of the set $\{1/m + 1/n : m, n \in \mathbf{N}\}$.
   Ans. $\left\{\frac{1}{n} : n \in \mathbf{N}\right\} \cup \{0\}$ or $\left\{\frac{1}{m} : m \in \mathbf{N}\right\} \cup \{0\}$ or
   $\left\{\frac{1}{m} : m \in \mathbf{N}\right\} \cup \{0\}$.
2. Find the limit points of the following sets:
   a. $\left\{\frac{n}{n+1} : n \in \mathbf{N}\right\}$.
   b. $\left\{1 + \frac{(-1)^n}{n} : n \in \mathbf{N}\right\}$.

c. $\left\{(-1)^n + \frac{1}{n} : n \in \mathbf{N}\right\}$.

d. $\left\{2^n + \frac{1}{2^n} : n \in \mathbf{N}\right\}$.

Ans. a. 1. b. 1. c. -1, 1. d. φ.

3. Find the limit points of the following set:

$$\left\{1, -1, 1\frac{1}{2}, -1\frac{1}{2}, 1\frac{1}{3}, -1\frac{1}{3}, \ldots\right\}.$$

Ans. 1 & –1.

4. Show that the set $S = \left\{1, \frac{1}{2}, \frac{1}{3}, \frac{1}{4}, \ldots\right\}$ is neither open nor closed.

5. Show that the set $S = \left\{1, -1, \frac{1}{2}, -\frac{1}{2}, \frac{1}{3}, \ldots\right\}$ is neither open nor closed.

6. Show that the set $S = \left\{1, -1, 1\frac{1}{2}, -1\frac{1}{2}, 1\frac{1}{3}, -1\frac{1}{3}, \ldots\right\}$ is closed, but not open.

7. Show the sets $\mathbf{N}$ and $\mathbf{Z}$ are not dense in $\mathbf{R}$.

Chapter 4

Sequences of Real Numbers

In this chapter, we discuss a very special class of functions whose domain is the set of natural numbers and the codomain is a subset of real numbers. Mainly, we will focus on real sequences. First, we introduce the meaning of convergence of these real sequences and then derive some basic results which are useful for the convergence of real sequences. Further, we present some important results for the convergence of sequences. These results include the Sandwich Theorem, the Monotone Convergence Theorem, the Cauchy First and Second Theorems, and the Ceasaro Theorem. It is useful to study these theorems and see how these theorems apply for special sequences. In continuation, we introduce the Cauchy sequences and derive some important results for them. At the end of this chapter, we present subsequences of real numbers.

4.1 Sequence

A real sequence is a function in which the domain is the set of natural numbers and the co-domain is the set of real numbers. Since, in the sequence, the domain of the function is the set of natural numbers, a sequence is completely determined if its values are given for all natural numbers. That is, $x(n)$ ($\forall\, n \in \mathbf{N}$) is given. The sequence $x(n)$ is denoted by $\langle x_n \rangle$ or $\{x_n\}_{n=1}^{\infty}$ or $\{x_n\}$ or $\langle x_1, x_2, x_3, \ldots, x_n, \ldots \rangle$. In a sequence, x_1 is the first and x_2 is the second, $\ldots$, and x_n is known as the nth term of the sequence.

Range of a sequence. Two terms of a sequence are distinct if $x_n \neq x_m$ for $n \neq m$. The collection of all distinct terms of a sequence is known as the range of a sequence.

Examples.

1. Sequence $\langle x_n \rangle = \langle (-1)^n \rangle = \langle -1, 1, -1, 1, \ldots \rangle$. The range of the sequence is a finite set $\{-1, 1\}$. So, the range of a sequence can be finite.
2. Sequence $\langle x_n \rangle = \langle (-1)^n + 1 \rangle = \langle 0, 2, 0, 2, \ldots \rangle$ and the range of the sequence is a finite set $\{0, 2\}$.
3. Sequence $\langle x_n \rangle = \langle \frac{1}{n} \rangle = \langle 1, \frac{1}{2}, \frac{1}{3}, \frac{1}{4}, \ldots \rangle$ and the range of the sequence is an infinite set $\{1, \frac{1}{2}, \frac{1}{3}, \frac{1}{4}, \ldots\}$.
4. Sequence $\langle x_n \rangle = \langle \frac{(-1)^n}{n^2} \rangle = \langle -1, \frac{1}{2^2}, -\frac{1}{3^2}, \frac{1}{4^2}, \ldots \rangle$ and the range of the sequence is an infinite set $\{-1, \frac{1}{4}, -\frac{1}{9}, \frac{1}{16}, \ldots\}$.

Bounded sequence. A sequence $\langle x_n \rangle$ is said to be bounded above if there exists a real number M such that $x_n \leq M$ ($\forall n \in \mathbf{N}$).

A sequence $\langle x_n \rangle$ is said to be bounded below if there exists a real number m such that

$$m \leq x_n \quad (\forall n \in \mathbf{N}).$$

A sequence $\langle x_n \rangle$ is said to be bounded if it is bounded above as well as bounded below. That is, there exist two real numbers m and M such that

$$m \leq x_n \leq M (\forall n \in \mathbf{N}).$$ We can also write $|x_n| \leq M$ ($\forall n \in \mathbf{N}$).

Simply, we can say that a sequence $\langle x_n \rangle$ is bounded if its range is bounded.

Unbounded sequence. A sequence $\langle x_n \rangle$ is unbounded if it is not bounded.

Examples.

1. Sequence $\langle (-1)^n \rangle = \langle -1, 1, -1, 1, \ldots \rangle$ and $-1 \leq x_n \leq 1, \forall n \in \mathbf{N}$. So, the sequence $\langle (-1)^n \rangle$ is bounded.
2. Sequence $\langle (-1)^n + 1 \rangle = \langle 0, 2, 0, 2, \ldots \rangle$ and $0 \leq x_n \leq 2, \forall n \in \mathbf{N}$. So, the sequence $\langle (-1)^n + 1 \rangle$ is bounded.
3. Sequence $\langle \frac{1}{n} \rangle = \langle 1, \frac{1}{2}, \frac{1}{3}, \frac{1}{4}, \ldots \rangle$ and $0 < x_n \leq 1, \forall n \in \mathbf{N}$. So, the sequence $\langle \frac{1}{n} \rangle$ is bounded.

4. Sequence $\langle n^2 \rangle = \langle 1, 4, 9, 16, \ldots \rangle$. This sequence is bounded below by 1, but not bounded above. So, this sequence is unbounded.
5. Sequence $\langle -n \rangle = \langle -1, -2, -3, -4, \ldots \rangle$. This sequence is bounded above by -1, but not bounded below. So, this sequence is unbounded.
6. Sequence $\langle (-1)^n n^2 \rangle = \langle -1, 4, -9, 16, \ldots \rangle$. This sequence is neither bounded below nor bounded above. So, this sequence is unbounded.

4.2 Convergence of Sequences

A sequence $\langle x_n \rangle$ is said to have a limit l if, for each $\varepsilon > 0$, there exists a natural number m depending on ε such that

$$|x_n - l| < \varepsilon, \quad \forall n \geq m.$$

A sequence $\langle x_n \rangle$ is said to be convergent if it has a finite limit l.

If a sequence $\langle x_n \rangle$ is said to be convergent to a finite limit l, then it means only a finite number of terms of the sequence lies outside the interval $(l - \varepsilon, l + \varepsilon)$. If a sequence $\langle x_n \rangle$ converges to a finite limit l, then we write $\lim_{n \to \infty} x_n = l$.

Theorem 1. *Every convergent sequence is bounded, but the converse is not true.*

Proof. Let $\langle x_n \rangle$ be a convergent sequence which converges to l. Then, for $\varepsilon > 0$, there exists a natural number m such that

$$|x_n - l| < \varepsilon, \quad \forall n \geq m.$$
$$\Longleftrightarrow x_n \in (l - \varepsilon, l + \varepsilon), \quad \forall n \geq m.$$
$$\Longleftrightarrow x_n < l + \varepsilon \quad \text{and} \quad l - \varepsilon < x_n, \quad \forall n \geq m.$$

Let $M = \max\{x_1, x_2, x_3, \ldots, x_{m-1}, l + \varepsilon\}$ and

$$m = \min\{x_1, x_2, x_3, \ldots, x_{m-1}, l - \varepsilon\}.$$

Then

$$m \leq x_n \leq M, \quad \forall n \in \mathbf{N}.$$

Hence, the sequence $\langle x_n \rangle$ is bounded.

To show that the converse is not true, we give an example of a sequence which is bounded, but not convergent. Consider the sequence $\langle (-1)^n \rangle = \langle -1, 1, -1, 1, \ldots \rangle$. Then $-1 \leq x_n \leq 1 \forall\ n \in \mathbf{N}$. So, the sequence $\langle (-1)^n \rangle$ is bounded. If possible, let $\lim_{n \to \infty} x_n = l$. Then, for $\varepsilon = \frac{1}{3}$, there exists a natural number m such that

$$|x_n - l| < \frac{1}{3}, \quad \forall n \geq m.$$

$$\Longleftrightarrow |(-1)^n - l| < \frac{1}{3}, \quad \forall\ n \geq m.$$

In particular, taking $n = 2m$ and $n = 2m + 1$, we have

$$|(-1)^{2m} - l| < \frac{1}{3} \quad \text{and} \quad |(-1)^{2m+1} - l| < \frac{1}{3}.$$

$$\Rightarrow |1 - l| < \frac{1}{3} \quad \text{and} \quad |-1 - l| < \frac{1}{3}.$$

$$\Rightarrow |1 - l| < \frac{1}{3} \quad \text{and} \quad |1 + l| < \frac{1}{3}.$$

Now, $2 = |1 - l + 1 + l| \leq |1 - l| + |1 + l| < \frac{1}{3} + \frac{1}{3} = \frac{2}{3}$.

$\Rightarrow 2 < \frac{2}{3}$, which is absurd. So, our assumption that the sequence $\langle (-1)^n \rangle$ converges to l is false. Hence, the sequence $\langle (-1)^n \rangle$ is not convergent.

Theorem 2. *A sequence of real numbers cannot converge to more than one limit.*

Proof. If possible, *let* $\langle x_n \rangle$ be a sequence which converges to two different limits l and l'.

Let the sequence $\langle x_n \rangle$ converge to l. Then, for $\varepsilon > 0$, there exists a natural number n_1 such that

$$|x_n - l| < \frac{\varepsilon}{2} \quad \text{for all } n \geq n_1.$$

Let the sequence $\langle x_n \rangle$ converge to l'. Then, for $\varepsilon > 0$, there exists a natural number n_2 such that

$$|x_n - l'| < \frac{\varepsilon}{2} \quad \text{for all } n \geq n_2.$$

Let $m = \max\{n_1, n_2\}$. Then

$$|x_n - l| < \frac{\varepsilon}{2}, \quad \text{and} \quad |x_n - l'| < \frac{\varepsilon}{2}, \quad \text{for all } n \geq m.$$

Let us assume that $\varepsilon = |l - l'| > 0$, because $l \neq l'$. Now, from the triangle inequality, we can write

$$|l - l'| = |l - x_n + x_n - l'| \leq |x_n - l| + |x_n - l'|,$$
$$< \frac{\varepsilon}{2} + \frac{\varepsilon}{2} = \varepsilon.$$

That is, $\varepsilon < \varepsilon$.

This is absurd. So, our assumption that a real sequence can converge to more than one limit is false, and hence, the limit of a real sequence is unique.

Example 1. $\lim \left(\frac{1}{n}\right) = 0$. For $\varepsilon = 0.1$ and 2, find the values of natural numbers after which this sequence will converge.

If $\varepsilon > 0$ is given, then $\frac{1}{\varepsilon} > 0$. By the Archimedean property, there is a natural number M, depending on ε, such that $\frac{1}{M} < \varepsilon$. So, if $n \geq M$, then we have $\frac{1}{n} \leq \frac{1}{M} < \varepsilon$, and hence, for $n \geq M$, we can write

$$\left|\frac{1}{n} - 0\right| = \frac{1}{n} < \varepsilon.$$

Therefore, the sequence $\langle \frac{1}{n} \rangle$ converges to 0.

Since the sequence converges for $M > \frac{1}{\varepsilon}$, for $\varepsilon = 0.1$, we have

$$M > \frac{1}{0.1} = 10, \quad \text{that is, } n = 11.$$

For $\varepsilon = 2$, we have

$$M > \frac{1}{2} = 0.5, \quad \text{that is, } n = 1.$$

Example 2. $\lim \left(\frac{4n + 3}{2n + 1}\right) = 2$.

Given $\varepsilon > 0$, we want to find a natural number M such that the following inequality holds true:

$$\left| \frac{4n+3}{2n+1} - 2 \right| < \varepsilon \quad \text{for all } n \geq M.$$

Simplifying the left-hand inequality, we have

$$\left| \frac{4n+3}{2n+1} - 2 \right| = \left| \frac{4n+3-4n-2}{2n+1} \right| = \left| \frac{1}{2n+1} \right| = \frac{1}{2n+1} < \frac{1}{n}. \quad (4.1)$$

Now, if the inequality $\frac{1}{n} < \varepsilon$ is satisfied, then the inequality in Eq. (4.1) is also satisfied. By the Archimedean property, there exists a natural number M, depending on ε, such that $\frac{1}{M} < \varepsilon$. So, if $n \geq M$, then we have $\frac{1}{n} \leq \frac{1}{M} < \varepsilon$ and hence, for $n \geq M$, we can write

$$\left| \frac{4n+3}{2n+1} - 2 \right| < \varepsilon.$$

Hence, the sequence $< \frac{4n+3}{2n+1} >$ converges to 2.

Example 3. If $0 < a < 1$, then $\lim(a^n) = 0$. Also find the value of the natural number after which this sequence converges if $a = 0.7$ and $\varepsilon = 0.002$.

For a given $\varepsilon > 0$, we want to find a natural number M such that the following inequality holds true:

$$|a^n - 0| < \varepsilon \quad \text{for all } n \geq M.$$

Simplifying the left-hand inequality, we have

$$|a^n - 0| = |a^n| = a^n < \varepsilon. \quad (4.2)$$

$$\Leftrightarrow n \log a < \log \varepsilon,$$

$$\Leftrightarrow n > \frac{\log \varepsilon}{\log a},$$

which is the reversed inequality because $\log a < 0$ for $0 < a < 1$.

Now, if the inequality $n > \frac{\log \varepsilon}{\log a}$ is satisfied, then the inequality in Eq. (4.2) is also satisfied. By the Archimedean property, there exists a natural number M, depending on ε, such that $M > \frac{\log \varepsilon}{\log a}$. So, if $n \geq M$, then we have $n > \frac{\log \varepsilon}{\log a}$ and hence, for $n \geq M$, we can write

$$|a^n - 0| < \varepsilon.$$

Hence, the sequence $\langle a^n \rangle$ for $0 < a < 1$ converges to 0.

Now, for $a = 0.7$ and $\varepsilon = 0.002$, we get

$$M > \frac{\log \varepsilon}{\log a} = \frac{\log 0.002}{\log 0.7} = 17.42373.$$

So, $M = 18$ is a suitable choice for $a = 0.7$ and $\varepsilon = 0.002$.

4.3 Limit Superior and Limit Inferior

Let $\langle x_n \rangle$ be a sequence of real numbers. Then the limit superior and the limit inferior are defined by

$$\lim_{n \to \infty} \sup x_n = \inf \sup \{x_n, x_{n+1}, x_{n+2}, \ldots\},$$

and

$$\lim_{n \to \infty} \inf x_n = \sup \inf \{x_n, x_{n+1}, x_{n+2}, \ldots\}.$$

The limit superior and the limit inferior of a sequence $\langle x_n \rangle$ are denoted by $\overline{\lim} x_n$ and $\underline{\lim} x_n$, respectively.

Remark. If $\langle x_n \rangle$ is any sequence, then we have the following relations between their limits:

$$\inf x_n \leq \underline{\lim} x_n \leq \overline{\lim} x_n \leq \sup x_n.$$

Example 1. Consider the sequence $x_n = (-1)^{n+1} (\forall n \in \mathbf{N})$.

$$\overline{\lim} x_n = \inf \sup \{(-1)^{n+1}, (-1)^{n+2}, (-1)^{n+3}, \ldots\}$$
$$= \lim_{n \to \infty} 1 = 1 (\forall\, n \in \mathbf{N})$$

and

$$\underline{\lim} x_n = \sup \inf \{(-1)^{n+1}, (-1)^{n+2}, (-1)^{n+3}, \ldots\} = \lim_{n \to \infty} -1$$
$$= -1 \ (\forall\, n \in \mathbf{N}).$$

Theorem 1. *If $\langle x_n \rangle$ and $\langle y_n \rangle$ are two sequences such that $\lim x_n = l$ and $\lim y_n = l'$, then*

I. $\lim(x_n + y_n) = \lim x_n + \lim y_n = l + l'$.

II. $\lim(x_n - y_n) = \lim x_n - \lim y_n = l - l'$.

III. $\lim(x_n y_n) = \lim x_n \lim y_n = ll'$.

IV. $\lim \left(\frac{x_n}{y_n} \right) = \frac{\lim x_n}{\lim y_n} = \frac{l}{l'}$, *if $l' \neq 0$ and $y_n \neq 0$ for all $n \in \mathbf{N}$.*

But the converse of the above theorem is not true.

Proof.

I. Let $\varepsilon > 0$ be given. Since $\lim x_n = l$ and $\lim y_n = l'$, there exist two natural numbers n_1 and n_2 such that

$$|x_n - l| < \frac{\varepsilon}{2} \quad \text{for all } n \geq n_1 \text{ and}$$

$$|y_n - l'| < \frac{\varepsilon}{2} \quad \text{for all } n \geq n_2.$$

Let $m = \max\{n_1, n_2\}$. Then

$$|x_n - l| < \frac{\varepsilon}{2} \quad \text{and} \quad |y_n - l'| < \frac{\varepsilon}{2} \quad \text{for all } n \geq m.$$

Now, we consider

$$|(x_n + y_n) - (l + l')| = |x_n - l + y_n - l'| \leq |x_n - l| + |y_n - l'|,$$

$$< \frac{\varepsilon}{2} + \frac{\varepsilon}{2} = \varepsilon.$$

$$\Rightarrow |(x_n + y_n) - (l + l')| < \varepsilon, \quad \text{for all } n \geq m.$$

Therefore, $\lim(x_n + y_n) = \lim x_n + \lim y_n = l + l'$.

II. Let $\varepsilon > 0$ be given. Since $\lim x_n = l$ and $\lim y_n = l'$, there exist two natural numbers n_1 and n_2 such that

$$|x_n - l| < \frac{\varepsilon}{2} \quad \text{for all } n \geq n_1 \text{ and}$$

$$|y_n - l'| < \frac{\varepsilon}{2} \quad \text{for all } n \geq n_2.$$

Let $m = \max\{n_1, n_2\}$. Then

$$|x_n - l| < \frac{\varepsilon}{2} \quad \text{and} \quad |y_n - l'| < \frac{\varepsilon}{2} \quad \text{for all } n \geq m.$$

Now, we get

$$|(x_n - y_n) - (l - l')| = |(x_n - l) - (y_n - l')|$$
$$\leq |x_n - l| + |y_n - l'|,$$
$$< \frac{\varepsilon}{2} + \frac{\varepsilon}{2} = \varepsilon.$$
$$\Rightarrow |(x_n - y_n) - (l - l')| < \varepsilon \quad \text{for all } n \geq m.$$

Therefore, we have

$$\lim(x_n - y_n) = \lim x_n - \lim y_n.$$

III. Consider

$$|(x_n y_n) - (ll')| = |x_n y_n - x_n l' + x_n l' - ll'|$$
$$= |x_n(y_n - l') + l'(x_n - l)|,$$
$$\leq |x_n||y_n - l'| + |l'||x_n - l|.$$

Since the sequences $\langle x_n \rangle$ and $\langle y_n \rangle$ are convergent, they are bounded. So, there exist M and M' such that

$$|x_n| \leq M \quad \text{and} \quad |y_n| \leq M' \quad \text{for all } n.$$

Therefore, we can write

$$|(x_n y_n) - (ll')| \leq M|y_n - l'| + |l'||x_n - l|.$$

Let $\varepsilon > 0$ be given. Since $\lim x_n = l$ and $\lim y_n = l'$, there exist two natural numbers n_1 and n_2 such that

$$|x_n - l| < \frac{\varepsilon}{2(|l'| + 1)} \quad \text{for all } n \geq n_1 \text{ and}$$

$$|y_n - l'| < \frac{\varepsilon}{2M} \quad \text{for all } n \geq n_2.$$

Let $m = \max\{n_1, n_2\}$. Then

$$|x_n - l| < \frac{\varepsilon}{2(|l'| + 1)} \quad \text{and} \quad |y_n - l'| < \frac{\varepsilon}{2M} \quad \text{for all } n \geq m.$$

Now, we consider

$$|(x_n y_n) - (ll')| \leq M|y_n - l'| + |l'||x_n - l|$$
$$< M\frac{\varepsilon}{2M} + |l'|\frac{\varepsilon}{2(|l'| + 1)} < \varepsilon.$$

$$|(x_n y_n) - (ll')| < \varepsilon \text{ for all } n \geq m.$$

Therefore, $\lim(x_n y_n) = \lim x_n \lim y_n = ll'$.

IV.
$$\left| \frac{x_n}{y_n} - \frac{l}{l'} \right| = \left| \frac{x_n l' - y_n l}{l' y_n} \right| = \left| \frac{x_n l' - l l' + l l' - y_n l}{l' y_n} \right|$$

$$= \left| \frac{l'(x_n - l) + l(y_n - l')}{l' y_n} \right|$$

$$\leq \frac{|l'||x_n - l| + |l||y_n - l'|}{|l'||y_n|}. \tag{4.3}$$

Now, $\lim y_n = l'$ and $l' \neq 0$. Let $\varepsilon = \frac{|l'|}{2} > 0$ be given. There exists a natural number n_1 such that

$$|y_n - l'| < \frac{|l'|}{2} \quad \text{for all } n \geq n_1. \tag{4.4}$$

Thus, for $n \geq n_1$, we have

$$||l'| - |y_n|| < |y_n - l'| < \frac{|l'|}{2}, \Rightarrow |y_n| \geq \frac{|l'|}{2}. \tag{4.5}$$

From Eqs. (4.3) and (4.4), we have

$$\left| \frac{x_n}{y_n} - \frac{l}{l'} \right| \leq \frac{2}{|l'|} |x_n - l| + \frac{2|l|}{|l'|^2} |y_n - l'|. \tag{4.6}$$

Let $\varepsilon > 0$ be given. There exist two natural numbers n_2 and n_3 such that

$$|x_n - l| < \frac{1}{4} |l'| \varepsilon \quad \text{for all } n \geq n_2 \tag{4.7}$$

and

$$|y_n - l'| < \frac{1}{4} \frac{|l'|^2}{(|l| + 1)} \varepsilon, \quad \text{for all } n \geq n_3. \tag{4.8}$$

Let $m = \max \{n_1, n_2, n_3\}$. Then, from Eqs. (4.7) and (4.8), we have

$$|x_n - l| < \frac{1}{4} |l'| \varepsilon \text{ and } |y_n - l'| < \frac{1}{4} \frac{|l'|^2}{(|l| + 1)} \varepsilon \text{ for all } n \geq m. \tag{4.9}$$

From Eqs. (4.6) and (4.7), we get

$$\left| \frac{x_n}{y_n} - \frac{l}{l'} \right| < \varepsilon \quad \text{for all } n \geq m.$$

Thus,

$$\lim \left(\frac{x_n}{y_n} \right) = \frac{\lim x_n}{\lim y_n} = \frac{l}{l'}.$$

Remark. To show that the converse is not true, we consider

$$x_n = (-1)^{n+1} \quad \text{and} \quad y_n = (-1)^n.$$

Then $x_n + y_n = 0$ for all $n \in \mathbf{N}$, $x_n y_n = -1$ for all $n \in \mathbf{N}$ and $\frac{x_n}{y_n} = -1$ for all $n \in \mathbf{N}$.

So, the sequence $\langle x_n + y_n \rangle$ converges to 0, the sequence $\langle x_n y_n \rangle$ converges to -1, and the sequence $\langle \frac{x_n}{y_n} \rangle$ converges to -1. But the sequences $\langle x_n \rangle$ and $\langle y_n \rangle$ are oscillatory, and hence, cannot be convergent.

Now, consider $x_n = (-1)^{n+1}$ and $y_n = (-1)^{n+1}$. Then $x_n - y_n = 0$ for all $n \in \mathbf{N}$.

So, the sequence $\langle x_n - y_n \rangle$ converges to 0, but the sequences $\langle x_n \rangle$ and $\langle y_n \rangle$ are oscillatory, and hence cannot be convergent.

Theorem 2. *Let $\langle x_n \rangle$ be a sequence of real numbers and let l be a real number. Also let $\langle b_n \rangle$ be a sequence of positive real numbers such that $\lim b_n = 0$. If, for some constants $M > 0$, and $m \in \mathbf{N}$,*

$$|x_n - l| \leq M b_n \quad (\forall\, n \geq m),$$

then $\lim x_n = l$.

Proof. Let $\varepsilon > 0$. Since $\lim b_n = 0$, then there exists a natural *number k* such that

$$|b_n - 0| < \frac{\varepsilon}{M} \quad (\forall\, n \geq k)$$

and

$$b_n < \frac{\varepsilon}{M} \quad (\forall\, n \geq k).$$

Given that

$$|x_n - l| \leq M b_n \quad (\forall\, n \geq m),$$

let $n_1 = \max\{k, m\}$. Then

$$b_n < \frac{\varepsilon}{M} \quad \text{and} \quad |x_n - l| \leq Mb_n \quad (\forall\, n \geq n_1)$$

and

$$|x_n - l| \leq Mb_n < M\frac{\varepsilon}{M} = \varepsilon \quad (\forall\, n \geq n_1).$$

That is, $|x_n - l| < \varepsilon \quad (\forall\, n \geq n_1)$.

Hence, $\lim x_n = l$.

Example 1. If $a > 0$, then $\lim a^{\frac{1}{n}} = 1$.

Solution. We consider the following three cases.

Case 1. If $a = 1$, then the sequence $\langle a^{\frac{1}{n}} \rangle$ is a constant sequence $\langle 1, 1, 1, \ldots \rangle$ and, therefore, $\lim a^{\frac{1}{n}} = 1$.

Case 2. If $a > 1$, then $a^{\frac{1}{n}} = 1 + b_n$ for some $b_n > 0$. Hence,

$$a = (1 + b_n)^n \geq 1 + nb_n \quad \text{for all } n \in \mathbf{N}.$$

$$\Rightarrow b_n \leq \frac{a - 1}{n} \quad \text{for all } n \in \mathbf{N}.$$

Now, $|a^{\frac{1}{n}} - 1| = |b_n| = b_n \leq (a - 1)\frac{1}{n}$ for all $n \in \mathbf{N}$.

By the previous theorem, taking $\lim \frac{1}{n} = 0$ and $M = a - 1 > 0$, we get

$$\lim a^{\frac{1}{n}} = 1.$$

Case 3. If $0 < a < 1$, then $a^{\frac{1}{n}} = \frac{1}{1+b_n}$ for some $b_n > 0$. Hence,

$$a = \frac{1}{(1 + b_n)^n} \leq \frac{1}{1 + nb_n} < \frac{1}{nb_n} \quad \text{for all } n \in \mathbf{N}.$$

$$\Rightarrow b_n < \frac{1}{na} \quad \text{for all } n \in \mathbf{N}.$$

Now, $|a^{\frac{1}{n}} - 1| = |\frac{b_n}{1+b_n}| < b_n \leq (\frac{1}{a})\frac{1}{n}$ for all $n \in \mathbf{N}$.

By the previous theorem again, taking $\lim \frac{1}{n} = 0$ and $M = \frac{1}{a} > 0$, we get

$$\lim a^{\frac{1}{n}} = 1.$$

Theorem 3. *Let* $\langle x_n \rangle$ *be a sequence of real numbers with* $x_n \geq 0 (\forall\ n \in \mathbf{N})$. *Then* $\lim x_n = l \geq 0$.

Proof. If possible, let $l < 0$. Then $\varepsilon = -l > 0$. Since $\lim x_n = l$, there exists a natural number m such that

$$|x_n - l| < \varepsilon \quad \text{for all } n \geq m.$$
$$\Rightarrow l - \varepsilon < x_n < l + \varepsilon \quad \text{for all } n \geq m.$$

In particular, taking $n = m$, then from the above inequality, we can write

$$x_m < l + \varepsilon = l - l = 0.$$
$$\Rightarrow x_m < 0,$$

which contradicts the fact that $x_n \geq 0\ \ \forall\ n \in \mathbf{N}$. So, our assumption is false, and hence $l \geq 0$.

Theorem 4 (Sandwich theorem). Let $\langle x_n \rangle, \langle y_n \rangle$, and $\langle z_n \rangle$ be three sequences such that $x_n \leq y_n \leq z_n$ $(\forall\ n \in \mathbf{N})$ and $\lim x_n = \lim z_n = l$. Then $\lim y_n = l$.

Proof. Let $\varepsilon > 0$ be given. Since $\lim x_n = l$ and $\lim z_n = l$, there exist two natural numbers n_1 and n_2 such that

$$|x_n - l| < \varepsilon \quad \text{for all } n \geq n_1$$

and

$$|z_n - l| < \varepsilon \quad \text{for all } n \geq n_2.$$

Let $m = \max\{n_1, n_2\}$. Then

$$|x_n - l| < \varepsilon \quad \text{and} \quad |z_n - l| < \varepsilon \quad \text{for all } n \geq m.$$

That is, $l - \varepsilon < x_n < l + \varepsilon$ and $l - \varepsilon < z_n < l + \varepsilon$ for all $n \geq m$.

$$\Rightarrow -\varepsilon < x_n - l < \varepsilon \quad \text{and} \quad -\varepsilon < z_n - l < \varepsilon \quad \text{for all } n \geq m. \quad (4.10)$$

Given that $x_n \leq y_n \leq z_n$ ($\forall\, n \in \mathbf{N}$).

$$\Rightarrow\ x_n - l \leq y_n - l \leq z_n - l\ (\forall\, n \in \mathbf{N}). \tag{4.11}$$

From Eqs. (4.10) and (4.11), we get

$$-\varepsilon < y_n - l < \varepsilon \quad \text{for all } n \geq m.$$
$$\Rightarrow l - \varepsilon < y_n < l + \varepsilon \quad \text{for all } n \geq m.$$
$$\Rightarrow |y_n - l| < \varepsilon, \quad \text{for all } n \geq m.$$

That is, $\lim y_n = l$.

Example 1. $\lim \left(\frac{\cos n}{n} \right) = 0$.

Proof. Since $-1 \leq \cos n \leq 1$, then

$$-\frac{1}{n} \leq \frac{\cos n}{n} \leq \frac{1}{n} \quad (\forall\, n \in \mathbf{N}).$$

We know that $\lim \frac{1}{n} = 0$ and $\lim \frac{-1}{n} = 0$.
Applying the Sandwich theorem, we get $\lim \left(\frac{\cos n}{n} \right) = 0$.

Theorem 5. *Let $\langle x_n \rangle$ be a sequence of real numbers such that $\lim \frac{x_{n+1}}{x_n} = l$. If $|l| < 1$, then the sequence $\langle x_n \rangle$ converges and $\lim x_n = 0$.*

Proof. Let k be a number such that $|l| < k < 1$. Then $\varepsilon = k - |l| > 0$. Since $\lim \frac{x_{n+1}}{x_n} = l$, there exists a natural number m such that

$$\left| \frac{x_{n+1}}{x_n} - l \right| < \varepsilon \quad \text{for all } n \geq m.$$

$$\Rightarrow \left| \frac{x_{n+1}}{x_n} \right| - |l| \leq \left| \frac{x_{n+1}}{x_n} - l \right| < \varepsilon \quad \text{for all } n \geq m.$$

$$\Rightarrow \left| \frac{x_{n+1}}{x_n} \right| < \varepsilon + |l| = k \quad \text{for all } n \geq m.$$

$$\Rightarrow \left| \frac{x_{n+1}}{x_n} \right| < k \quad \text{for all } n \geq m.$$

$$\Rightarrow |x_{n+1}| < k|x_n| < k^2|x_{n-1}| < k^3|x_{n-2}| < k^4|x_{n-3}|$$
$$< \cdots < k^{n-m+1}|x_m| \quad \text{for all } n \geq m.$$
$$\Rightarrow k|x_n| < k^{n-m+1}|x_m| \quad \text{for all } n \geq m.$$
$$\Rightarrow |x_n| < k^n \frac{|x_m|}{k^m} \quad \text{for all } n \geq m.$$

Let $M = \frac{|x_m|}{k^m}$

$$\Rightarrow |x_n| < k^n M \quad \text{for all } n \geq m.$$

Since $0 < k < 1$ and since we know that $\lim k^n = 0$, if $0 < k < 1$, therefore, $\lim x_n = 0$.

Example 1. $\lim \frac{n+1}{3^{n+1}} = 0$.

Proof. Let $x_n = \frac{n+1}{3^{n+1}}$. Then $x_{n+1} = \frac{n+2}{3^{n+2}}$.

$$\lim \frac{x_{n+1}}{x_n} = \lim \left(\frac{(n+2)(3^{n+1})}{(n+1)(3^{n+2})} \right) = \frac{1}{3} \lim \left(\frac{\left(1 + \frac{2}{n}\right)}{\left(1 + \frac{1}{n}\right)} \right) = \frac{1}{3} < 1.$$

Applying Theorem 5, $\lim \frac{n+1}{3^{n+1}} = 0$.

Theorem 6. *Let* $\langle x_n \rangle$ *be a sequence of real numbers such that* $\lim \frac{x_{n+1}}{x_n} = l$. *If* $l > 1$, *then the sequence* $\langle x_n \rangle$ *diverges and* $\lim x_n = \infty$.

Proof. Let k be a number such that $1 < k < l$. Then $\varepsilon = l - k > 0$. Since $\lim \frac{x_{n+1}}{x_n} = l$, there exists a natural number m such that

$$\left| \frac{x_{n+1}}{x_n} - l \right| < \varepsilon \quad \text{for all } n \geq m.$$

$$\Rightarrow l - \varepsilon < \frac{x_{n+1}}{x_n} < l + \varepsilon \quad \text{for all } n \geq m.$$

$$\Rightarrow \frac{x_{n+1}}{x_n} > l - \varepsilon = k \quad \text{for all } n \geq m.$$

$$\Rightarrow \frac{x_{n+1}}{x_n} > k \quad \text{for all } n \geq m.$$

$$\Rightarrow x_{n+1} > kx_n > k^2 x_{n-1} > k^3 x_{n-2} > k^4 x_{n-3}$$

$$> \cdots > k^{n-m+1} x_m \quad \text{for all } n \geq m.$$
$$\Rightarrow k x_n > k^{n-m+1} x_m \quad \text{for all } n \geq m.$$
$$\Rightarrow x_n > k^n \frac{x_m}{k^m} \quad \text{for all } n \geq m.$$

Let $M = \dfrac{x_m}{k^m}$.

$$\Rightarrow x_n > k^n M \quad \text{for all } n \geq m.$$

We know that $\lim k^n = \infty$, if $k > 1$. Therefore, $\lim x_n = \infty$.

Example 1. $\lim \frac{2^n}{n} = \infty$.

Proof. Let $x_n = \dfrac{2^n}{n}$. Then $x_{n+1} = \dfrac{2^{n+1}}{n+1}$.

$$\lim \frac{x_{n+1}}{x_n} = \lim \left(\frac{(n)(2^{n+1})}{(n+1)(2^n)} \right) = 2 \lim \left(\frac{1}{\left(1 + \frac{1}{n}\right)} \right) = 2 > 1.$$

Applying Theorem 6, we obtain $\lim \frac{2^n}{n} = \infty$.

4.4   Monotone Sequences

4.4.1   *Monotonic increasing sequences*

A sequence $\langle x_n \rangle$ is said to be a monotonic increasing sequence if it satisfies the following condition:

$$x_1 \leq x_2 \leq x_3 \leq \cdots \leq x_n \leq x_{n+1} \leq \cdots$$

and it is said to be a strictly monotonic increasing sequence if it satisfies the following condition:

$$x_1 < x_2 < x_3 < \cdots < x_n < x_{n+1} < \cdots .$$

4.4.2 *Monotonic decreasing sequences*

A sequence $\langle x_n \rangle$ is said to be a monotonic decreasing sequence if it satisfies the following condition:

$$x_1 \geq x_2 \geq x_3 \geq \cdots \geq x_n \geq x_{n+1} \geq \cdots$$

and it is said to be a strictly monotonic decreasing sequence if it satisfies the following condition:

$$x_1 > x_2 > x_3 > \cdots > x_n > x_{n+1} > \cdots .$$

A sequence is said to be monotone if it is either increasing or decreasing.

Examples.

I. $\langle 1, 1, 2, 2, 3, 3, \ldots \rangle$ is a monotonic increasing sequence.
II. $\langle 1, 2, 3, 4, \ldots \rangle$ is a strictly monotonic increasing sequence.
III. $\langle -1, -1, -2, -2, -3, -3, \ldots \rangle$ is a monotonic decreasing sequence.
IV. $\langle 1, \frac{1}{2}, \frac{1}{3}, \ldots, \frac{1}{n}, \frac{1}{n+1}, \ldots \rangle$ is a strictly monotonic decreasing sequence.
V. $\langle 1, -2, 3, -4, 5, -6, \ldots \rangle$ is not a monotone sequence.

All of the above discussed theorems on limits can be applied if the limit is known. Now, we will give some theorems based on monotonic sequences to calculate the limit of a sequence.

Theorem 1. *A bounded monotonic increasing sequence is convergent and it converges to its supremum.*

Proof. Let $\langle x_n \rangle$ be a bounded monotonic increasing sequence. Since $\langle x_n \rangle$ is bounded, there exists a real number M such that

$$x_n \leq M \quad (\forall\, n \in \mathbf{N}).$$

Now, apply the completeness property of real-number sequences that, if a sequence $\langle x_n \rangle$ is bounded above, then it has supremum in $\mathbf{R}$. Let it be l. Now, we will show that the sequence $\langle x_n \rangle$ converges to l.

Thus, whenever $n \geq m$, we have $x_m \leq x_n$, because $\langle x_n \rangle$ is a monotonic increasing sequence.

Since l is the supremum of the set $\{x_n : n \in \mathbf{N}\}$, by the property of supremum, there exists a member x_m such that

$$l - \varepsilon < x_m.$$

Combining the above two equations, we can write

$$l - \varepsilon < x_m \leq x_n \leq l < l + \varepsilon \quad \text{whenever } n \geq m.$$
$$\Rightarrow l - \varepsilon < x_n < l + \varepsilon \quad \text{whenever } n \geq m.$$
$$\Rightarrow |x_n - l| < \varepsilon \quad \text{whenever } n \geq m.$$

Therefore, the sequence $\langle x_n \rangle$ is convergent and it converges to its supremum. That is, $\lim x_n = l$.

Theorem 2. *A bounded monotonic decreasing sequence is convergent and it converges to its infimum.*

Proof. Let $\langle x_n \rangle$ be a bounded monotonic decreasing sequence. Since $\langle x_n \rangle$ is bounded, there exists a real number M such that

$$x_n \leq M \quad (\forall \, n \in \mathbf{N}).$$

Now, apply the completeness property of real number that, if a sequence $\langle x_n \rangle$ is bounded below, then it has infimum in $\mathbf{R}$. Let it be l. Now, we will show that the sequence $\langle x_n \rangle$ converges to l.

If, whenever $n \geq m$, we have
$x_n \leq x_m$ because $\langle x_n \rangle$ is a monotonic decreasing sequence.

Now, since l is the infimum of the set $\{x_n : n \in \mathbf{N}\}$, by the property of infimum, there exists a member x_m such that

$$l + \varepsilon > x_m.$$

Combining the above two equations, we can write

$$l + \varepsilon > x_m \geq x_n \geq l > l - \varepsilon \quad \text{whenever } n \geq m.$$
$$\Rightarrow l - \varepsilon < x_n < l + \varepsilon \quad \text{whenever } n \geq m.$$
$$\Rightarrow |x_n - l| < \varepsilon \quad \text{whenever } n \geq m.$$

So, the sequence $\langle x_n \rangle$ is convergent and it converges to its infimum. That is, $\lim x_n = l$.

Example 1. Show that the sequence $\langle x_n \rangle$, defined by the recursion formula $x_{n+1} = \sqrt{5x_n}$, with $x_1 = 1$, converges to 5.

Solution. First, we will show that the sequence $\langle x_n \rangle$ is bounded by the principle of mathematical induction. It is true for $n = 1$,

$$x_1 = 1 < 5.$$

Let it be true for n. That is, $x_n < 5$. Then we will prove it to be true for $n + 1$. That is, $x_n < 5$,

$$\Rightarrow \sqrt{x_n} < \sqrt{5},$$
$$\Rightarrow \sqrt{5x_n} < \sqrt{5 \times 5},$$
$$\Rightarrow x_{n+1} < 5.$$

So, by the principle of mathematical induction, the given sequence is bounded by 5.

Now, we will prove that the given sequence is a monotonic increasing sequence. It is true for $n = 1$:

$x_1 < x_2$. Let it be true for n. That is, $x_n < x_{n+1}$,

$$\Rightarrow \sqrt{5x_n} < \sqrt{5x_{n+1}},$$
$$\Rightarrow x_{n+1} < x_{n+2}.$$

So, it is true for $n + 1$. Hence, the given sequence is a monotonic increasing sequence. Since the given sequence is a bounded monotonic increasing sequence, it is convergent. Let it converge to l. That is, $\lim x_n = l = \lim x_{n+1}$. Then

$$\lim x_{n+1} = \sqrt{5 \lim x_n},$$
$$\Rightarrow l = \sqrt{5l},$$
$$\Rightarrow l = 0, 5.$$

Since $x_n > 1$ ($\forall\, n \in \mathbf{N}$), it cannot converge to 0. Therefore, $l = 5$.

Example 2. Show that the sequence $\langle x_n \rangle$, defined by the recursion formula $x_{n+1} = \frac{1}{4}(2x_n + 3)$ for $n \geq 1$ and $x_1 = 1$, converges to $\frac{3}{2}$.

Solution. First, we will show that the sequence $\langle x_n \rangle$ is bounded by the principle of mathematical induction. It is true for $n = 1$.

$$x_1 < x_2 = \frac{5}{4} < 2.$$

Let it be true for n. That is, $x_n < 2$. Then we will prove it for $n + 1$.

$$x_{n+1} = \frac{1}{4}(2x_n + 3) < \frac{1}{4}(2 \times 2 + 3) = \frac{7}{4} < 2,$$

$$\Rightarrow x_{n+1} < 2.$$

So, by the principle of mathematical induction, the given sequence is bounded by 2.

Now, we will prove that the given sequence is a monotonic increasing sequence. It is true for $n = 1$

$x_1 < x_2$. Let it be true for n. That is, $x_n < x_{n+1}$, so that

$$x_{n+1} = \frac{1}{4}(2x_n + 3) < \frac{1}{4}(2x_{n+1} + 3) = x_{n+2}.$$

$$\Rightarrow x_{n+1} < x_{n+2}.$$

So, it is true for $n + 1$. Hence, the given sequence is a monotonic increasing sequence. Since the given sequence is a bounded monotonic increasing sequence, it is convergent. Let it converge to l. That is, $\lim x_n = l = \lim x_{n+1}$. Then

$$\lim x_{n+1} = \frac{1}{4}(2 \lim x_n + 3),$$

$$\Rightarrow l = \frac{1}{4}(2l + 3),$$

$$\Rightarrow l = \frac{3}{2}.$$

4.5   Some Important Theorems on Limits

4.5.1   *Cauchy's first theorem on limits*

Let $\langle x_n \rangle$ be a sequence such that $\lim x_n = l$. Then

$$\lim \frac{x_1 + x_2 + x_3 + \cdots + x_n}{n} = l.$$

Proof. First, we define a new sequence $\langle c_n \rangle$, which is given as $c_n = x_n - l$. Then

$$\lim c_n = \lim x_n - l = l - l = 0.$$

Now,

$$\lim \frac{c_1 + c_2 + c_3 + \cdots + c_n}{n}$$

$$= \lim \frac{(x_1 - l) + (x_2 - l) + (x_3 - l) + \cdots + (x_n - l)}{n}$$

$$= \lim \frac{x_1 + x_2 + x_3 + \cdots + x_n}{n} - l.$$

$$\Rightarrow \lim \frac{x_1 + x_2 + x_3 + \cdots + x_n}{n}$$

$$= \lim \frac{c_1 + c_2 + c_3 + \cdots + c_n}{n} + l.$$

Now, if we will prove that $\lim \frac{c_1 + c_2 + c_3 + \cdots + c_n}{n} = 0$, then our proof is complete.

Since the sequence $\langle x_n \rangle$ is convergent, it is bounded, and hence the sequence $\langle c_n \rangle$ is bounded. Therefore, there exists a constant $K > 0$, such that

$$|c_n| \leq K \quad \text{for all } n \in \mathbf{N}. \tag{4.12}$$

Since $\lim c_n = 0$, there exists a natural number m_1 such that

$$|c_n| < \frac{\varepsilon}{2} \quad \text{for all } n \geq m_1. \tag{4.13}$$

From Eq. (4.12), we can write

$$\left| \frac{c_1 + c_2 + c_3 + \cdots + c_{m_1}}{n} \right| \leq \frac{1}{n}(|c_1| + |c_2| + |c_3| + \cdots + |c_{m_1}|)$$

$$\leq \frac{m_1 K}{n} \quad \text{for all } n \in \mathbf{N}. \tag{4.14}$$

For any $\varepsilon > 0$, we can always find a natural number m_2 such that

$$\frac{m_1 K}{n} < \frac{\varepsilon}{2} \quad \text{for all } n \geq m_2. \tag{4.15}$$

From Eqs. (4.14) and (4.15), we can conclude that

$$\left| \frac{c_1 + c_2 + c_3 + \cdots + c_{m_1}}{n} \right| < \frac{\varepsilon}{2} \quad \text{for all } n \geq m_2 \tag{4.16}$$

From Eq. (4.13), we can write

$$\left| \frac{c_{m_1+1} + c_{m_1+2} + c_{m_1+3} + \cdots + c_n}{n} \right|$$

$$\leq \frac{1}{n} (|c_{m_1+1}| + |c_{m_1+2}| + |c_{m_1+3}| + \cdots |c_n|)$$

$$< \frac{\varepsilon}{2} \frac{(n - m_{1)}}{n}, \left| \frac{c_{m_1+1} + c_{m_1+2} + c_{m_1+3} + \cdots + c_n}{n} \right|$$

$$< \frac{\varepsilon}{2} \frac{(n - m_{1)}}{n} \quad \text{for all } n \geq m_1. \tag{4.17}$$

Let $m = \max\{m_1, m_2\}$. Then

$$\left| \frac{c_1 + c_2 + c_3 + \cdots + c_{m_1}}{n} \right| < \frac{\varepsilon}{2} \quad \text{for all } n \geq m \tag{4.18}$$

$$\left| \frac{c_{m_1+1} + c_{m_1+2} + c_{m_1+3} + \cdots + c_n}{n} \right| < \frac{\varepsilon}{2} \frac{(n - m_{1)}}{n} \quad \text{for all } n \geq m. \tag{4.19}$$

Now, from Eqs. (4.18) and (4.19), we conclude that

$$\left| \frac{c_1 + c_2 + c_3 + \cdots + c_n}{n} - 0 \right|$$

$$= \left| \frac{c_1 + c_2 + c_3 + \cdots + c_n}{n} \right| \leq \left| \frac{c_1 + c_2 + c_3 + \cdots + c_{m_1}}{n} \right|$$

$$+ \left| \frac{c_{m_1+1} + c_{m_1+2} + c_{m_1+3} + \cdots + c_n}{n} \right|$$

$$< \frac{\varepsilon}{2} + \frac{\varepsilon}{2} \frac{(n - m_{1)}}{n} \quad \text{for all } n \geq m,$$

$$< \frac{\varepsilon}{2} + \frac{\varepsilon}{2} = \varepsilon \quad \text{for all } n \geq m, \quad \text{because } \frac{(n - m_{1)}}{n} < 1$$

$$\left| \frac{c_1 + c_2 + c_3 + \cdots + c_n}{n} - 0 \right| < \varepsilon \quad \text{for all } n \geq m.$$

$$\Rightarrow \lim \frac{c_1 + c_2 + c_3 + \cdots + c_n}{n} = 0.$$

$$\Rightarrow \lim \frac{x_1 + x_2 + x_3 + \cdots + x_n}{n} = l.$$

Example 1. Evaluate $\lim \dfrac{\left(\frac{1}{1^2} + \frac{1}{2^2} + \cdots \frac{1}{n^2} \right)}{n}$.

Solution. Let $x_n = \frac{1}{n^2}$. Then $\lim \frac{1}{n^2} = 0$.
Using the Cauchy first theorem on limits, we get

$$\lim \frac{\left(\frac{1}{1^2} + \frac{1}{2^2} + \cdots \frac{1}{n^2} \right)}{n} = 0.$$

Example 2. Find $\lim \left[\dfrac{1}{\sqrt{n^2+1}} + \dfrac{1}{\sqrt{n^2+2}} + \cdots \dfrac{1}{\sqrt{n^2+n}} \right]$.

Solution. We can write

$$\left[\frac{1}{\sqrt{n^2+1}} + \frac{1}{\sqrt{n^2+2}} + \cdots \frac{1}{\sqrt{n^2+n}} \right]$$

$$= \frac{1}{n} \left[\frac{1}{\sqrt{1+\frac{1}{n^2}}} + \frac{1}{\sqrt{1+\frac{2}{n^2}}} + \cdots \frac{1}{\sqrt{1+\frac{n}{n^2}}} \right].$$

Let $x_k = \dfrac{1}{\sqrt{1+\frac{k}{n^2}}}$ $(k = 1, 2, \ldots, n)$. Then

$$\lim x_n = \lim \frac{1}{\sqrt{1+\frac{n}{n^2}}} = \lim \frac{1}{\sqrt{1+\frac{1}{n}}} = 1.$$

Using the Cauchy first theorem on limits, we get

$$\lim \left[\frac{1}{\sqrt{n^2+1}} + \frac{1}{\sqrt{n^2+2}} + \cdots \frac{1}{\sqrt{n^2+n}} \right]$$

$$= \lim \frac{1}{n} \left[\frac{1}{\sqrt{1+\frac{1}{n^2}}} + \frac{1}{\sqrt{1+\frac{2}{n^2}}} + \cdots \frac{1}{\sqrt{1+\frac{n}{n^2}}} \right] = 1.$$

Example 3. Find $\lim \left[\dfrac{1}{\sqrt{3n^2+1}} + \dfrac{1}{\sqrt{3n^2+2}} + \cdots + \dfrac{1}{\sqrt{3n^2+n}} \right].$

Solution. Let $x_k = \dfrac{1}{\sqrt{3+\frac{k}{n^2}}}$ $(k = 1, 2, \ldots, n)$.

Then $\lim x_n = \lim \dfrac{1}{\sqrt{3+\frac{n}{n^2}}} = \lim \dfrac{1}{\sqrt{3+\frac{1}{n}}} = \dfrac{1}{\sqrt{3}}$. Using the Cauchy first theorem on limits, we get

$$\lim \left[\frac{1}{\sqrt{3n^2+1}} + \frac{1}{\sqrt{3n^2+2}} + \cdots \frac{1}{\sqrt{3n^2+n}} \right]$$

$$= \lim \frac{1}{n} \left[\frac{1}{\sqrt{3+\frac{1}{n^2}}} + \frac{1}{\sqrt{3+\frac{2}{n^2}}} + \cdots \frac{1}{\sqrt{3+\frac{n}{n^2}}} \right] = \frac{1}{\sqrt{3}}.$$

Example 4. Evaluate $\lim \dfrac{1}{n} \left[1 + 2^{\frac{1}{2}} + 3^{\frac{1}{3}} + \cdots + n^{\frac{1}{n}} \right].$

Solution. Let $x_n = n^{\frac{1}{n}}$. Then $\lim x_n = \lim n^{\frac{1}{n}} = 1$. Using the Cauchy first theorem on limits, we get

$$\lim \frac{1}{n} \left[1 + 2^{\frac{1}{2}} + 3^{\frac{1}{3}} + \cdots + n^{\frac{1}{n}} \right] = \lim n^{\frac{1}{n}} = 1.$$

4.5.2 Cauchy's second theorem on limits

Let $\langle x_n \rangle$ be a sequence of positive numbers such that $\lim x_n = l$. Then

$$\lim (x_1, x_2, x_3, \ldots, x_n)^{1/n} = l.$$

Proof. Let $\lim x_n = l$.

Since $\langle x_n \rangle$ is a sequence of positive numbers, we can define a new sequence $\langle y_n \rangle$ such that

$$y_n = \log x_n \quad \text{for all } n \in \mathbf{N}.$$

Then

$$\lim y_n = \lim \log x_n = \log \lim x_n = \log l.$$

Using the Cauchy first theorem on limits, we can write

$$\lim \frac{y_1 + y_2 + y_3 + \cdots + y_n}{n} = \log l,$$

$$\lim \frac{\log x_1 + \log x_2 + \log x_3 + \cdots + \log x_n}{n} = \log l,$$

$$\lim \frac{\log(x_1 x_2 x_3 \cdots x_n)}{n} = \log l,$$

$$\lim \log(x_1 x_2 x_3 \cdots x_n)^{1/n} = \log l,$$

$$\Rightarrow \log \lim(x_1 x_2 x_3 \cdots x_n)^{1/n} = \log l,$$

$$\Rightarrow \lim(x_1 x_2 x_3 \cdots x_n)^{1/n} = l.$$

Example 1. Find $\lim \left[\frac{2}{1} \left(\frac{3}{2}\right)^2 \left(\frac{4}{3}\right)^3 \cdots \left(\frac{n+1}{n}\right)^n \right]^{\frac{1}{n}}$.

Solution. Let $x_n = \left(\frac{n+1}{n}\right)^n = \left(n + \frac{1}{n}\right)^n$. Then $x_n > 0$ and

$$\lim x_n = \lim \left(n + \frac{1}{n}\right)^n = e.$$

Using the Cauchy second theorem on limits, we get

$$\lim(x_1 x_2 x_3 \cdots x_n)^{1/n} = e.$$

$$\Rightarrow \lim \left[\frac{2}{1} \left(\frac{3}{2}\right)^2 \left(\frac{4}{3}\right)^3 \cdots \left(\frac{n+1}{n}\right)^n \right]^{\frac{1}{n}} = e.$$

Example 2. Find $\lim \left[1.2^{\frac{1}{2}}.3^{\frac{1}{3}} \cdots n^{\frac{1}{n}} \right]^{\frac{1}{n}}$

Solution. Let $x_n = n^{\frac{1}{n}}$. Then $x_n > 0$ and $\lim x_n = \lim n^{\frac{1}{n}} = 1$.
Using the Cauchy second theorem on limits, we get

$$\lim \left[1.2^{\frac{1}{2}}.3^{\frac{1}{3}} \cdots n^{\frac{1}{n}} \right]^{\frac{1}{n}} = \lim n^{\frac{1}{n}} = 1.$$

4.5.3   Cesaro's theorem

If $\langle x_n \rangle$ and $\langle y_n \rangle$ are two convergent sequences such that $\lim x_n = l$ and $\lim y_n = l'$, then

$$\lim \frac{x_1 y_n + x_2 y_{n-1} + \cdots + x_n y_1}{n} = ll'.$$

Proof.   Let $x_n = l + c_n$, where $\lim |c_n| = 0$.

Since the sequence $\langle |c_n| \rangle$ converges to 0, by the Cauchy first theorem on limits, we can write

$$\lim \frac{|c_1| + |c_2| + \cdots + |c_n|}{n} = 0. \tag{4.20}$$

Since $\lim y_n = l'$, by the Cauchy first theorem on limits, we can write

$$\lim \frac{y_1 + y_2 + \cdots + y_n}{n} = l'. \tag{4.21}$$

Since every convergent sequence of real numbers is also bounded, the sequence $\langle y_n \rangle$ is bounded. So, there exists a real number $B > 0$ such that

$$|y_n| \leq B \quad \text{for all } n \in \mathbf{N}. \tag{4.22}$$

$$\lim \frac{x_1 y_n + x_2 y_{n-1} + \cdots + x_n y_1}{n}$$

$$= \lim \frac{(l + c_1)y_n + (l + c_2)y_{n-1} + \cdots + (l + c_n)y_1}{n}$$

$$= \lim \frac{l(y_1 + y_2 + \cdots + y_n)}{n} + \lim \frac{c_1 y_n + c_2 y_{n-1} + \cdots + c_n y_1}{n}. \tag{4.23}$$

From Eq. (4.21), we get

$$\lim \frac{l(y_1 + y_2 + \cdots + y_n)}{n} = l.$$

$$\lim \frac{(y_1 + y_2 + \cdots + y_n)}{n} = ll'. \tag{4.24}$$

From Eq. (4.22), we can write

$$\left| \frac{c_1 y_n + c_2 y_{n-1} + \cdots + c_n y_1}{n} \right| \leq \frac{1}{n}(|c_1 y_n| + |c_2 y_{n-1}| + \cdots |c_n y_1|)$$

$$\leq \frac{B}{n}(|c_1| + |c_2| + \cdots |c_n|),$$

$$0 \leq \left| \frac{c_1 y_n + c_2 y_{n-1} + \cdots + c_n y_1}{n} \right|$$

$$\leq \frac{B}{n}(|c_1| + |c_2| + \cdots |c_n|). \quad (4.25)$$

Using Eq. (4.20), and the sandwich theorem in Eq. (4.25), we get

$$\lim \frac{c_1 y_n + c_2 y_{n-1} + \cdots + c_n y_1}{n} = 0. \quad (4.26)$$

From Eqs. (4.23), (4.24), and (4.26), we get

$$\lim \frac{x_1 y_n + x_2 y_{n-1} + \cdots + x_n y_1}{n} = ll'.$$

4.6 Cauchy Sequences

A sequence $\langle x_n \rangle$ is said to be a Cauchy sequence if, for a given $\varepsilon > 0$, there exists a natural number m such that

$$|x_n - x_m| < \varepsilon \quad \text{whenever } n \geq m.$$

Equivalently, a sequence $\langle x_n \rangle$ is said to be a Cauchy sequence if, for a given $\varepsilon > 0$, there exists a natural number p such that

$$|x_n - x_m| < \varepsilon \quad \text{whenever } n, m \geq p.$$

This means that a sequence is a Cauchy sequence if its terms come close to each other after a finite number of terms.

Example 1. The sequence $\langle \frac{1}{n} \rangle$ is a Cauchy sequence.

Solution. Let $\varepsilon > 0$ be given and $x_n = \frac{1}{n}$. Then, for $n > m$,

$$|x_n - x_m| = \left| \frac{1}{n} - \frac{1}{m} \right| \leq \left| \frac{1}{n} \right| + \left| \frac{1}{m} \right| = \frac{1}{n} + \frac{1}{m}.$$

By the Archimedean property, we can find a natural number p such that $p > \frac{2}{\varepsilon}$. Then, for $n, m \geq p$, we have

$$\frac{1}{n}, \frac{1}{m} \leq \frac{1}{p} < \frac{\varepsilon}{2},$$

$$\Rightarrow \frac{1}{n} < \frac{\varepsilon}{2} \quad \text{and} \quad \frac{1}{m} < \frac{\varepsilon}{2}.$$

$$\Rightarrow |x_n - x_m| \leq \frac{1}{n} + \frac{1}{m} < \frac{\varepsilon}{2} + \frac{\varepsilon}{2} = \varepsilon.$$

So for $n, m \geq p$, we get

$$|x_n - x_m| < \varepsilon.$$

Therefore, $\langle \frac{1}{n} \rangle$ is a Cauchy sequence.

Example 2. The sequence $\langle \frac{(-1)^n}{n} \rangle$ is a Cauchy sequence.

Solution. Let $\varepsilon > 0$ be given and $x_n = \frac{(-1)^n}{n}$. Then, for $n > m$,

$$|x_n - x_m| = \left| \frac{(-1)^n}{n} - \frac{(-1)^m}{m} \right| \leq \left| \frac{(-1)^n}{n} \right| + \left| \frac{(-1)^m}{m} \right| = \frac{1}{n} + \frac{1}{m}.$$

By the Archimedean property, we can find a natural number p such that $p > \frac{2}{\varepsilon}$. Then, for $n, m \geq p$, $|x_n - x_m| \leq \frac{1}{n} + \frac{1}{m} < \frac{\varepsilon}{2} + \frac{\varepsilon}{2} = \varepsilon$.
So, for $n, m \geq p$, we have

$$|x_n - x_m| < \varepsilon.$$

Therefore, $\langle \frac{(-1)^n}{n} \rangle$ is a Cauchy sequence.

Example 3. The sequence $\langle x_n \rangle$, where $x_n = 1 + \frac{1}{2} + \frac{1}{3} + \cdots + \frac{1}{n}$ is not a Cauchy sequence.

Solution. For $n > m$,

$$|x_n - x_m| = \left| \left(1 + \frac{1}{2} + \frac{1}{3} + \cdots \frac{1}{m} + \frac{1}{m+1} + \cdots + \frac{1}{n} \right) \right.$$

$$\left. - \left(1 + \frac{1}{2} + \frac{1}{3} + \cdots + \frac{1}{m} \right) \right|$$

$$= \left| \frac{1}{m+1} + \frac{1}{m+2} + \frac{1}{m+3} + \cdots \frac{1}{n} \right|$$

$$= \frac{1}{m+1} + \frac{1}{m+2} + \frac{1}{m+3} + \cdots \frac{1}{n}.$$

In particular, taking $n = 2m$, we get

$$|x_{2m} - x_m| = \frac{1}{m+1} + \frac{1}{m+2} + \frac{1}{m+3} + \cdots \frac{1}{2m}$$

$$\geq \frac{1}{2m} + \frac{1}{2m} + \frac{1}{2m} + \cdots \frac{1}{2m}$$

$$= m \times \frac{1}{2m} = \frac{1}{2}, |x_{2m} - x_m| > \frac{1}{2}. \tag{4.27}$$

If $\langle x_n \rangle$ is a Cauchy sequence, then, for all $n > m$ and $\varepsilon > 0$, we have

$$|x_n - x_m| < \varepsilon.$$

But, for $\varepsilon = \frac{1}{3}$ and $n = 2m$, and from Eq. (4.27), we get

$$|x_{2m} - x_m| \not< \frac{1}{3}.$$

Therefore, $\langle x_n \rangle$ is not a Cauchy sequence.

Example 4. The sequence $\langle x_n \rangle$, where $x_n = 1 + \frac{1}{4} + \frac{1}{7} + \cdots + \frac{1}{(3n-2)}$ is not a Cauchy sequence.

Solution. For $n > m$,

$$|x_n - x_m| = \left| \left(1 + \frac{1}{4} + \frac{1}{7} + \cdots \frac{1}{3m-2} + \frac{1}{3m+1} + \cdots + \frac{1}{3n-2} \right) \right.$$

$$\left. - \left(1 + \frac{1}{4} + \frac{1}{7} + \cdots + \frac{1}{3m-2} \right) \right|$$

$$= \left| \frac{1}{3m+1} + \frac{1}{3m+4} + \cdots \frac{1}{3n-2} \right|$$

$$= \frac{1}{3m+1} + \frac{1}{3m+4} + \cdots \frac{1}{3n-2}.$$

In particular, taking $n = 2m$, we get

$$
\begin{aligned}
|x_{2m} - x_m| &= \frac{1}{3m+1} + \frac{1}{3m+4} + \cdots \frac{1}{3m+3m-2} \\
&\geq \frac{1}{3m+3m} + \frac{1}{3m+3m} + \cdots \frac{1}{3m+3m} \\
&= m \times \frac{1}{6m} = \frac{1}{6}, |x_{2m} - x_m| > \frac{1}{6}.
\end{aligned}
\tag{4.28}
$$

If $\langle x_n \rangle$ is a Cauchy sequence, then, for all $n > m$ and $\varepsilon > 0$, we have

$$
|x_n - x_m| < \varepsilon.
$$

But, for $\varepsilon = \frac{1}{7}$ and $n = 2m$, and from Eq. (4.28), we get

$$
|x_{2m} - x_m| \nless \frac{1}{7}.
$$

Therefore, $\langle x_n \rangle$ is not a Cauchy sequence.

4.7 Subsequences

Let $\langle x_n \rangle$ be a sequence. Then a sequence formed by picking the terms from the sequence $\langle x_n \rangle$ in any way, but preserving the original order of terms as it was in the sequence $\langle x_n \rangle$, is called a subsequence. If $n_1 < n_2 < n_3 < \ldots$, then the sequence $\langle x_{n_k} \rangle$ is a subsequence of $\langle x_n \rangle$.

Example 1. $\{x_1, x_3, x_5, \ldots\}$ and $\{x_2, x_4, x_6, \ldots\}$ are subsequences of $\{x_1, x_2, x_3, x_4, x_5, \ldots\}$.

Example 2. $\{1, 1, 1, \ldots\}$ and $\{-1, -1, -1, \ldots\}$ are subsequences of $\{(-1)^n\}$.

Example 3. $\{\frac{1}{5}, \frac{1}{6}, \frac{1}{7}, \ldots\}$ is a subsequence of $\{\frac{1}{n}\}$ obtained by removing some finite number of terms.

Remark. Every bounded sequence has a convergent subsequence.

Theorem 1. *A real sequence $\langle x_n \rangle$ is convergent if and only if it is a Cauchy sequence.*

Proof. Let $\langle x_n \rangle$ be a convergent sequence and converge to a limit l. Then we will prove that $\langle x_n \rangle$ is a Cauchy sequence. Let $\varepsilon > 0$ be given. Then there exists a natural number m such that

$$|x_n - l| < \frac{\varepsilon}{2} \quad (\forall\, n \geq m).$$

In particular, we can write

$$|x_m - l| < \frac{\varepsilon}{2}.$$

Now, for $\varepsilon > 0$ and $n \geq m$, we have

$$|x_n - x_m| = |x_n - l - x_m + l| \leq |x_n - l| + |x_m - l| < \frac{\varepsilon}{2} + \frac{\varepsilon}{2} = \varepsilon,$$

$$\Rightarrow |x_n - x_m| < \varepsilon \quad (\forall\, n \geq m).$$

Therefore, $\langle x_n \rangle$ is a Cauchy sequence.

To prove the converse part, let $\langle x_n \rangle$ be a Cauchy sequence. Then we will prove that it is convergent. Every Cauchy sequence is bounded and every bounded sequence has a convergent subsequence. Let $\langle x_{n_k} \rangle$ be a convergent subsequence of *the sequence* $\langle x_n \rangle$ and suppose that it converges to a finite number l. Then, for $\varepsilon > 0$, there exists a natural number p_1 such that

$$|x_{n_k} - l| < \frac{\varepsilon}{2} \quad (\forall\, n_k \geq p_1). \tag{4.29}$$

We can also write

$$|x_B - l| < \frac{\varepsilon}{2} \quad (\forall\, B \geq p_1 \text{ and } B \in \{n_1, n_2, n_3, \ldots\}). \tag{4.30}$$

Since $\langle x_n \rangle$ is a Cauchy sequence, for $\varepsilon > 0$, there exists a natural number p_2 such that

$$|x_n - x_m| < \frac{\varepsilon}{2} \quad (\forall\, n, m \geq p_2). \tag{4.31}$$

Let $p = \max\{p_1, p_2\}$. Then, from Eqs. (4.30) and (4.31), we can write

$$|x_B - l| < \frac{\varepsilon}{2} \quad (\forall\, B \geq p) \tag{4.32}$$

and

$$|x_n - x_m| < \frac{\varepsilon}{2} \quad (\forall \ n, m \geq p). \tag{4.33}$$

Since $B \geq p$, in particular, by taking $m = B$ in Eq. (4.33), we get

$$|x_n - x_B| < \frac{\varepsilon}{2} \quad (\forall \ n \geq p).$$

Therefore, for $n \geq p$, we have

$$|x_n - l| = |x_n - l - x_B + x_B| \leq |x_n - x_B| + |x_B - l| < \frac{\varepsilon}{2} + \frac{\varepsilon}{2} = \varepsilon,$$

$$\Rightarrow |x_n - l| < \varepsilon \ (\forall \ n \geq p).$$

Therefore, $\langle x_n \rangle$ is a convergent sequence.

Exercises

1. Show that $\lim_{n \to \infty} \dfrac{1}{3^n} = 0$.

2. Show that $\lim_{n \to \infty} \dfrac{n^2 + 3n + 5}{2n^2 + 5n + 7} = \dfrac{1}{2}$.

3. Show that $\lim_{n \to \infty} \dfrac{2n + 3}{3n + 4} = \dfrac{2}{3}$.

4. Show that $\lim_{n \to \infty} \dfrac{n^2 + 1}{3n^2 + 2} = \dfrac{1}{3}$.

5. Show that the sequence $\langle x_n \rangle$, defined by the recursion formula $x_{n+1} = \sqrt{3x_n}$, $x_1 = 1$, converges to 3.

6. Show that the sequence $\langle x_n \rangle$, defined by the recursion formula $x_{n+1} = \sqrt{7x_n}$, $x_1 = 1$, converges to 7.

7. Show that the sequence $\langle x_n \rangle$, defined by the recursion formula $x_{n+1} = \sqrt{3 + x_n}$, $x_1 = \sqrt{3}$, converges to the positive root of $x^2 - x - 3 = 0$.

8. Show that the sequence $\langle x_n \rangle$, defined by the recursion formula $x_{n+1} = 2 - \frac{1}{x_n}$, for $n \geq 1$ and $x_1 = \frac{3}{2}$, converges to 1.

9. Find $\lim_{n \to \infty} \left[\frac{1}{\sqrt{5n^2+1}} + \frac{1}{\sqrt{5n^2+2}} + \cdots \frac{1}{\sqrt{5n^2+n}} \right]$.

   Ans $\frac{1}{\sqrt{5}}$.

10. Show that, if $x_n = \frac{1}{n}\{(n + 1)(n + 2)\cdots(n + n)\}^{1/n}$, then the sequence $\langle x_n \rangle$ converges to $\frac{4}{e}$.

11. Show that $\lim_{n \to \infty} \frac{n^2}{2^n} = 0$.

12. Show that the sequence $\langle x_n \rangle$, where $x_n = 1 + \frac{1}{3} + \frac{1}{5} + \cdots + \frac{1}{2n-1}$ is not a Cauchy sequence.

13. Show that the sequence $\langle x_n \rangle$, where $x_n = 1 + \frac{1}{1!} + \frac{1}{2!} + \cdots + \frac{1}{n!}$, is a Cauchy sequence.

Chapter 5

Limits and Continuity

The concept of limits is very important in Real Analysis. In Chapter 4, we have studied the concept of limits for real sequences. In this chapter, we will study the concept of limits for real functions. For the limit behavior of sequences, we always consider $n \to \infty$. But, in the limit behavior of functions, we consider $x \to b$, where b can be any finite or infinite real numbers.

Further, we study the most important class of functions that arises in real analysis: the class of continuous functions. First, we define the continuity of functions at a point and on a set. Then we define the discontinuity of functions and the types of discontinuity of functions. In continuation, we prove that the sum, difference, product, and quotient of two continuous functions is also a continuous function. We extend our study and present the concept of sequential continuity and, by using this concept, we will discuss the continuity of a most important function known as Dirichlet's function. In the end, we establish some important properties of continuous functions such as continuous functions on closed and bounded intervals must attain their minimum and maximum values, intermediate-value property, and fixed point property. These properties are not possessed by general functions, which is why the class of continuous functions is a very important class of functions in Real Analysis.

5.1   Limit of a Function

To calculate the limit of a function at a point, it is not necessary that the function should be defined at that point, but it should be defined in the neighborhood of the given point.

A number l is said to be the limit of a function f at a point b if for each $\varepsilon > 0$ there exists a $\delta > 0$ depending on ε such that

$$|f(x) - l| < \varepsilon \quad \text{whenever} \quad 0 < |x - b| < \delta.$$

Symbolically, it is written as follows:

$$\lim_{x \to b} f(x) = l.$$

In the definition of limits, we have taken $0 < |x - b|$. If $|x - b| = 0$, then $x = b$. But, for limits, it is not necessary that the function should be defined at that point. So, $0 < |x - b|$

$$\Longleftrightarrow \quad x \neq b.$$

To calculate the limit of a function at a point, we see the behavior of the function in the neighborhood of that point. When we approach the point b along the x-axis, then the function approaches l on the y-axis. To approach a point on the x-axis, there are only two possibilities. First, if we approach from the right to that point and second, if we approach from the left to that point. So, a function on the real line has two side limits called the right-hand limit and the left-hand limit.

Right-hand limit. If we approach the point b on the x-axis from the right, then the value of the function on the y-axis is known as the right-hand limit of a function at the point b. If l is the right-hand limit of a function f at a point b, then, for each $\varepsilon > 0$, there exists a $\delta > 0$ such that

$$|f(x) - l| < \varepsilon \quad \text{whenever} \quad b < x < b + \delta.$$

Symbolically, we write

$$\lim_{x \to b+0} f(x) = l, \quad \text{or} \quad f(b + 0) = l.$$

To calculate the right-hand limit, we put $x = b + h (h > 0)$ in the function $f(x)$ and take the limit as $h \to 0+$.

Right-hand limit $= $ R.H.L. $= f(b + 0) = \lim\limits_{x \to b+0} f(x) = \lim\limits_{h \to 0+} f(b + h)$.

Left-hand limit. If we approach the point b on the x-axis from the left, then the value of the function on the y axis is known as left-hand limit of the function at the point b. If l is the left-hand limit of a function f at a point b, then, for each $\varepsilon > 0$, there exists a $\delta > 0$ such that

$$|f(x) - l| < \varepsilon \quad \text{whenever} \quad b - \delta < x < b.$$

Symbolically, we write

$$\lim\limits_{x \to b-0} f(x) = l, \quad \text{or} \quad f(b - 0) = l.$$

To calculate the left-hand limit, put $x = b - h \ (h > 0)$ in the function $f(x)$ and take the limit as $h \to 0+$.

Left-hand limit $=$ L.H.L. $= f(b - 0) = \lim\limits_{x \to b-0} f(x) = \lim\limits_{h \to 0+} f(b - h)$.

We say that the limit of a function exists at a point b if both-side limits of the function exist and are equal. That is,

$$f(b + 0) = f(b - 0) \quad \text{or} \quad \lim\limits_{h \to 0+} f(b + h) = \lim\limits_{h \to 0+} f(b - h).$$

Example 1. Let $f(x) = x \ (\forall x \in \mathbf{R})$. Then $\lim_{x \to b} f(x) = b$.

Solution. Let $\varepsilon > 0$. Then

$$|f(x) - b| = |x - b|.$$

Choose $\delta = \varepsilon$. Then, if $0 < |x - b| < \delta$, we have

$$|f(x) - b| < \varepsilon.$$

So, for each $\varepsilon > 0$, we can find a $\delta = \varepsilon > 0$ such that

$$|f(x) - b| < \varepsilon \quad \text{whenever} \quad 0 < |x - b| < \delta.$$

Therefore, $\lim\limits_{x \to b} f(x) = b$.

Example 2. Let $f(x) = x^3$ ($\forall\, x \in \mathbf{R}$). Then $\lim_{x \to b} f(x) = b^3$.

Solution. Let $\varepsilon > 0$. Then

$$\left|f(x) - b^3\right| = \left|x^3 - b^3\right| = \left|(x - b)(x^2 + b^2 + xb)\right|$$
$$< |x - b|\,\left|x^2 + b^2 + xb\right|.$$

Let $|x - b| < 1$. Then $|x| < 1 + |b|$,

$$\Rightarrow \left|x^2 + b^2 + xb\right| < |x|^2 + |b|^2 + |xb| < (1 + |b|)^2 + |b|^2 + |b|(1 + |b|)$$
$$= 1 + 3\,|b|^2 + 3\,|b|$$
$$\left|x^2 + b^2 + xb\right| < 1 + 3|b|^2 + 3|b|.$$

Therefore, if $|x - b| < 1$, then we have

$$\left|x^3 - b^3\right| < (1 + 3|b|^2 + 3|b|)(|x - b|). \qquad (5.1)$$

We want to make this term less than ε. For doing this, we take $|x-b| < \frac{\varepsilon}{(1+3|b|^2+3|b|)}$. Therefore, we choose $\delta = \inf\left\{1,\ \frac{\varepsilon}{(1+3|b|^2+3|b|)}\right\}$. If $0 < |x - b| < \delta$, then it follows that $|x - b| < 1$. So, Eq. (5.1) is true and

$$\left|x^3 - b^3\right| < (1 + 3\,|b|^2 + 3\,|b|)\,(|x - b|) < \varepsilon.$$

So, for each $\varepsilon > 0$, we can find a $\delta(\varepsilon) > 0$ such that

$$\left|f(x) - b^3\right| < \varepsilon \quad \text{whenever } 0 < |x - b| < \delta.$$

Therefore, $\lim_{x \to b} f(x) = b^3$.

Example 3. Let $f(x) = \frac{x^2-4}{x^2+1}$ ($\forall\, x \in \mathbf{R}$). Then $\lim_{x \to 2} f(x) = 0$.

Solution. Let $\varepsilon > 0$. Then

$$|f(x) - 0| = \left|\frac{x^2 - 4}{x^2 + 1} - 0\right| = \left|\frac{(x - 2)(x + 2)}{x^2 + 1}\right| < |x - 2|\,\frac{|x + 2|}{x^2 + 1}.$$

Let $|x - 2| < 1$. Then $1 < |x| < 3$,

$$\Rightarrow\ |x + 2| < |x| + 2 < 5 \quad \text{and} \quad x^2 + 1 > 2.$$

Therefore, if $|x - 2| < 1$, then we have

$$\left|\frac{x^2 - 4}{x^2 + 1} - 0\right| < \frac{5}{2}\,|x - 2| \qquad (5.2)$$

We want to make this term less than ε. For doing this, we take $|x-2| < \frac{2}{5}\varepsilon$. Therefore, we choose $\delta = \inf\left\{1,\ \frac{2}{5}\varepsilon\right\}$. If $0 < |x - 2| < \delta$, then it follows that $|x - 2| < 1$. So, Eq. (5.2) is true and

$$\left| \frac{x^2 - 4}{x^2 + 1} - 0 \right| < \frac{5}{2} |x - 2| < \varepsilon.$$

So, for each $\varepsilon > 0$, we can find a $\delta(\varepsilon) > 0$, such that

$$|f(x) - 0| < \varepsilon \quad \text{whenever} \quad 0 < |x - 2| < \delta.$$

Therefore, $\lim_{x \to 2} f(x) = 0$.

Example 4. Let $f(x) = x \sin \frac{1}{x}$ ($\forall\, x \in \mathbf{R}$). Then $\lim_{x \to 0} f(x) = 0$.

Solution. Let $\varepsilon > 0$. Then

$$|f(x) - 0| = \left| x \sin \frac{1}{x} - 0 \right| = \left| x \sin \frac{1}{x} \right| \le |x| \quad \text{because} \quad \left| \sin \frac{1}{x} \right| \le 1.$$

Choose $\delta = \varepsilon$. If $0 < |x - 0| < \delta$, then we have

$$|f(x) - 0| < \varepsilon.$$

So, for each $\varepsilon > 0$, we can find a $\delta = \varepsilon > 0$ such that

$$\left| x \sin \frac{1}{x} - 0 \right| < \varepsilon \quad \text{whenever} \quad 0 < |x - 0| < \delta.$$

Therefore, $\lim_{x \to 0} f(x) = 0$.

5.2   Continuity at Interior Points

Let $f : [a, b] \to \mathbf{R}$. Then a function f is said to be continuous at an interior point c that is, $a < c < b$ if, for each $\varepsilon > 0$, there corresponds a $\delta > 0$ such that

$$|f(x) - f(c)| < \varepsilon \quad \text{whenever} \quad |x - c| < \delta. \tag{5.3}$$

The definition of continuity is similar to the definition of limits, but the only difference is that, in the case of limit, we take ($|f(x) - l| < \varepsilon$), whereas, in the case of continuity, we take ($|f(x) - f(c)| < \varepsilon$). That is, in the case of continuity, the function should be defined at that point. Now, using the definition of limits in Eq. (5.3), we can write

$$\lim_{x \to c} f(x) = f(c).$$

So, a function f is continuous at an interior point if the limit of the function exists at that point, and also if it is equal to the value of the function at that point.

5.2.1   *Continuity at boundary points*

A function $f : [a, b] \to \mathbf{R}$ is continuous at the point a if, for each $\varepsilon > 0$, there corresponds a $\delta > 0$ such that

$$|f(x) - f(a)| < \varepsilon \quad \text{whenever } a < x < a + \delta. \qquad (5.4)$$

Now, using the definition of limits in Eq. (5.4), we can write

$$\lim_{x \to a+0} f(x) = f(a).$$

So, a function f is continuous at the end point a if the right-hand limit of the function exists, and also if it is equal to the value of the function at that point.

A function $f : [a, b] \to \mathbf{R}$ is continuous at the point b if, for each $\varepsilon > 0$, there corresponds a $\delta > 0$ such that

$$|f(x) - f(b)| < \varepsilon \quad \text{whenever } b - \delta < x < b. \qquad (5.5)$$

Now, using the definition of limits in Eq. (5.5), we can write

$$\lim_{x \to b-0} f(x) = f(b).$$

So, a function f is continuous at the end point b if the left-hand limit of the function exists, and also if it is equal to the value of the function at that point.

Remarks.

1. A function f is continuous in an interval if it is continuous at each point of the interval.
2. A function f is continuous in a domain if it is continuous at each point of the domain.

Example 1. Show that the following function f is continuous at $x = 0$:

$$f(x) = \begin{cases} x \sin\left(\frac{1}{x}\right), & x \neq 0 \\ 0, & x = 0. \end{cases}$$

Solution. Here

$$\text{L. H. L. } f(0-0) = \lim_{h \to 0} f(0-h) = \lim_{h \to 0} \left(-h \sin\left(\frac{1}{-h}\right)\right)$$

$$= 0 \times \lim_{h \to 0} \left(\sin\left(\frac{1}{h}\right)\right) = 0 \times l = 0,$$

$$\text{because } \lim_{h \to 0} \sin\left(\frac{1}{h}\right) = l,$$

and l is a finite number lying between -1 to 1.

$$\text{R. H. L. } f(0+0) = \lim_{h \to 0} f(0+h) = \lim_{h \to 0} \left(h \sin\left(\frac{1}{h}\right)\right) = 0 \times l = 0.$$

Since $f(0-0) = f(0+0) = f(0) = 0$, therefore, the function f is continuous at $x = 0$.

Example 2. Show that the following function f is continuous everywhere:

$$f(x) = \begin{cases} -x, & x < 0, \\ 0, & 0 \leq x \leq 1, \\ 1 - x, & x > 1. \end{cases}$$

Solution. Given function is continuous everywhere except at the break points $x = 0$ and 1, where, we will check its continuity.

At $x = 0$

$$\text{L. H. L. } f(0-0) = \lim_{h \to 0} f(0-h) = \lim_{h \to 0}(h) = 0,$$

$$\text{R. H. L. } f(0+0) = \lim_{h \to 0} f(0+h) = \lim_{h \to 0}(0) = 0,$$

and $f(0) = 0.$

Since $f(0-0) = f(0+0) = f(0) = 0$, the function f is continuous at $x = 0$.

At $x = 1$

$$\text{L. H. L. } f(1-0) = \lim_{h\to0} f(1-h) = \lim_{h\to0}(0) = 0,$$

$$\text{R. H. L. } f(1+0) = \lim_{h\to0} f(1+h) = \lim_{h\to0}(1-(1+h))$$
$$= \lim_{h\to0}(-h) = 0,$$

and $f(1) = 0$.

Since $f(1-0) = f(1+0) = f(1) = 0$, the function f is continuous at $x = 1$.

So, the given function is continuous everywhere.

Example 3. Show that the function $f(x) = |x-1| + |x-2|$ is continuous everywhere.

Solution. Using the property of modulus, we can redefine the above function as

$$f(x) = \begin{cases} 3-2x, & x < 1, \\ 1, & 1 \leq x < 2, \\ 2x-3, & x \geq 2. \end{cases}$$

The given function is continuous everywhere except at the break points $x = 1$ and 2, where we will check its continuity.

At $x = 1$

$$\text{L. H. L. } f(1-0) = \lim_{h\to0} f(1-h) = \lim_{h\to0}(3-2(1-h))$$
$$= \lim_{h\to0}(1+2h) = 1.$$

$$\text{R. H. L. } f(1+0) = \lim_{h\to0} f(1+h) = \lim_{h\to0}(1) = 1,$$

and $f(1) = 1$.

Since $f(1-0) = f(1+0) = f(1) = 1$, the function f is continuous at $x = 1$.

At $x = 2$

L. H. L. $f(2-0) = \lim\limits_{h\to 0} f(2-h) = \lim\limits_{h\to 0}(1) = 1$

R. H. L. $f(2+0) = \lim\limits_{h\to 0} f(2+h) = \lim\limits_{h\to 0}(2(2+h) - 3)$

$$= \lim\limits_{h\to 0}(2h+1) = 1,$$

and $f(2) = 2 \times 2 - 3 = 1$.

Since $f(2-0) = f(2+0) = f(2) = 1$, the function f is continuous at $x = 2$.

So, the given function is continuous everywhere.

Example 4. Find the value of the constants m and n for which the following function f is continuous everywhere:

$$f(x) = \begin{cases} 3x - 2, & x \le 0, \\ mx^2 + 5n, & 0 < x < 1, \\ 3x + 2m, & x \ge 1. \end{cases}$$

Solution. Since the given function is continuous everywhere, it will be continuous at the break points $x = 0$ and 1.

At $x = 0$.

L. H. L. $f(0-0) = \lim\limits_{h\to 0} f(0-h) = \lim\limits_{h\to 0}(-3h - 2) = -2$.

R. H. L. $f(0+0) = \lim\limits_{h\to 0} f(0+h) = \lim\limits_{h\to 0}(mh^2 + 5n) = 5n$,

and $f(0) = 3 \times 0 - 2 = -2$.

Since f is continuous at $x = 0$, therefore, $f(0-0) = f(0+0) = f(0)$.

$$\Rightarrow 5n = -2,$$

$$\Rightarrow n = -\frac{2}{5}.$$

At $x = 1$.

L. H. L. $f(1-0) = \lim\limits_{h\to 0} f(1-h) = \lim\limits_{h\to 0}\left(m(1-h)^2 + 5n\right) = m + 5n$.

R. H. L. $f(1+0) = \lim\limits_{h\to 0} f(1+h) = \lim\limits_{h\to 0}(3(1+h) + 2m) = 2m + 3$,

and $f(1) = 3 \times 1 + 2m = 3 + 2m$.

Since f is continuous at $x = 1$, therefore, $f(1 - 0) = f(1 + 0) = f(1)$.

$$\Rightarrow m + 5n = 2m + 3,$$

$$\Rightarrow m = -5.$$

Therefore, the given function is continuous everywhere when $m = -5$ and $n = -\frac{2}{5}$.

5.3   Discontinuity of a Function

A function is said to be discontinuous at a point of its domain if it is not continuous at that point. Basically, there are only two possibilities for a function to be discontinuous at a point.

1. The limit of the function exists at the point, but it is not equal to the value of the function at that point.
2. The limit of the function does not exist at that point. If the limit of the function does not exist, then there are the following possibilities:

   a. Both the left-hand and the right-hand limits exist, but they are not equal to each other.
   b. Both the side limits do not exist.
   c. One side limit exists and the other side limit does not exist.

On the basis of the above observations, the discontinuity of a function at a point can be divided into the following parts.

1. **Removable discontinuity.** A function f is said to have a removable discontinuity at a point b if the limit of the function exists at b, but it is not equal to the value of the function at the point b. That is, $f(b - 0) = f(b + 0) \neq f(b)$.

This discontinuity is called removable discontinuity, because it can be removed by assigning $f(b)$ the same value as the value of limit at point b.

Example 1. Consider the function f defined by

$$f(x) = \begin{cases} \dfrac{\sin 2x}{x}, & x \neq 0 \\ 1, & x = 0. \end{cases}$$

Then L. H. L. $f(0-0) = \lim_{h \to 0} f(0-h) = \lim_{h \to 0} \dfrac{\sin(-2h)}{-h}$

$$= \lim_{h \to 0} \dfrac{\sin 2h}{h} = 2.$$

R. H. L. $f(0+0) = \lim_{h \to 0} f(0+h) = \lim_{h \to 0} \dfrac{\sin 2h}{h} = 2.$

Since $f(0-0) = f(0+0) = 2 \neq f(0) = 1$, therefore, function f has removable discontinuity at $x = 0$. Since this type of discontinuity can be removed, if we take $f(0) = 2$ and redefine given function as

$$f(x) = \begin{cases} \dfrac{\sin 2x}{x}, & x \neq 0 \\ 2, & x = 0 \end{cases}$$

then $f(0-0) = f(0+0) = f(0) = 2$. Now, this new function is continuous at $x = 0$.

2. **First-kind discontinuity.** A function f is said to have first-kind discontinuity at a point b if both side limits of the function exist at b, but they are not equal to each other. That is, $f(b-0)$ and $f(b+0)$ both exist, but $f(b-0) \neq f(b+0)$. This discontinuity is also called jump discontinuity, because, there is a jump in the function due to which both limits become different.

Example 1. Consider the function f defined by

$$f(x) = \begin{cases} -2, & x < 1 \\ 2, & x > 1. \end{cases}$$

Then L. H. L. $f(1-0) = \lim_{h \to 0} f(1-h) = \lim_{h \to 0} (-2) = -2.$

R. H. L. $f(1+0) = \lim_{h \to 0} f(1+h) = \lim_{h \to 0} (2) = 2.$

Since $f(1-0) \neq f(1+0)$, function f has first-kind discontinuity at $x = 1$.

Example 2. Consider the function f defined by

$$f(x) = \begin{cases} \dfrac{e^{\frac{1}{x}} - e^{-\frac{1}{x}}}{e^{\frac{1}{x}} + e^{-\frac{1}{x}}}, & x \neq 0 \\ 1, & x = 0. \end{cases}$$

Then L. H. L. $f(0-0) = \lim_{h \to 0} f(0-h) = \lim_{h \to 0} \left(\dfrac{e^{-\frac{1}{h}} - e^{\frac{1}{h}}}{e^{-\frac{1}{h}} + e^{\frac{1}{h}}} \right)$

$$= \lim_{h \to 0} \left(\dfrac{e^{\frac{1}{h}} \left(e^{-\frac{2}{h}} - 1 \right)}{e^{\frac{1}{h}} \left(e^{-\frac{2}{h}} + 1 \right)} \right) = \lim_{h \to 0} \left(\dfrac{e^{-\frac{2}{h}} - 1}{e^{-\frac{2}{h}} + 1} \right)$$

$$= \dfrac{0-1}{0+1} = -1, \text{ because } \lim_{h \to 0} \left(e^{-\frac{2}{h}} \right) = 0.$$

R. H. L. $f(0+0) = \lim_{h \to 0} f(0+h) = \lim_{h \to 0} \left(\dfrac{e^{\frac{1}{h}} - e^{-\frac{1}{h}}}{e^{\frac{1}{h}} + e^{-\frac{1}{h}}} \right)$

$$= \lim_{h \to 0} \left(\dfrac{e^{\frac{1}{h}} \left(1 - e^{-\frac{2}{h}} \right)}{e^{\frac{1}{h}} \left(1 + e^{-\frac{2}{h}} \right)} \right)$$

$$= \lim_{h \to 0} \left(\dfrac{1 - e^{-\frac{2}{h}}}{1 + e^{-\frac{2}{h}}} \right) = \dfrac{1-0}{1+0} = 1.$$

Since $f(0-0) \neq f(0+0)$, function f has first-kind discontinuity at $x = 0$.

3. **Second-kind discontinuity.** A function f is said to have second-kind discontinuity at a point b if both side limits of the function do not exist at b. That is, neither $f(b-0)$ nor $f(b+0)$ exist.

Example 1. Consider the function f defined by

$$f(x) = \begin{cases} \cos\left(\dfrac{1}{x} \right), & x \neq 0 \\ 1, & x = 0. \end{cases}$$

Then L. H. L. $f(0-0) = \lim_{h \to 0} f(0-h) = \lim_{h \to 0} \cos\left(\dfrac{1}{-h} \right) = a$ finite number between –1 to 1.

So, the left-hand limit does not exist.

R. H. L. $f(0 + 0) = \lim_{h \to 0} f(0 + h) = \lim_{h \to 0} \cos\left(\frac{1}{h}\right)$ = a finite number between -1 to 1.

So, the right-hand limit does not exist.

Since both the side limits $f(0 - 0)$ and $f(0 + 0)$ do not exist, the function f has the second-kind discontinuity at $x = 0$.

4. **Mixed discontinuity.** A function f is said to have mixed discontinuity at a point b if one side limit of the function exists at b, but the other side limit does not exist. That is, either $f(b - 0)$ or $f(b + 0)$ does not exist.

This type of discontinuity is called mixed discontinuity, because it is the combination of the first-kind and the second-kind of discontinuities. That is, one side limit exists, but the other side limit does not exist.

Example 1. Consider the function f defined by

$$f(x) = \begin{cases} \sin\left(\dfrac{1}{x}\right), & x < 0 \\ 1, & x \geq 0. \end{cases}$$

Then L. H. L. $f(0 - 0) = \lim_{h \to 0} f(0 - h) = \lim_{h \to 0} \sin\left(\frac{1}{-h}\right)$ = a finite number between -1 to 1.

So, the left-hand limit does not exist.

R. H. L. $f(0 + 0) = \lim_{h \to 0} f(0 + h) = \lim_{h \to 0}(1) = 1$.

Since the limit $f(0 - 0)$ does not exist and $f(0 + 0)$ exists, the function f has mixed discontinuity at $x = 0$. Since $f(0+0) = f(0) = 1$, the function is continuous from the right.

5. **Infinite discontinuity.** A function f is said to have an infinite discontinuity at a point b if one or more of the functional limits $\overline{f(b - 0)}$, $\underline{f(b - 0)}$, $\overline{f(b + 0)}$, and $\underline{f(b + 0)}$ are ∞ or $-\infty$.

Example 1. Discuss the continuity of the function $f(x) = x - \text{sgn}(x)$, where $\text{sgn}(x)$ is the signum function defined as

$$\text{sgn}(x) = \begin{cases} -1, & x < 0, \\ 0, & x = 0, \\ 1, & x > 0. \end{cases}$$

Solution. Using the property of the signum function, we can redefine the above function as follows:

$$f(x) = \begin{cases} x - 1, & x < 0, \\ 0, & x = 0, \\ x + 1, & x > 0. \end{cases}$$

The given function is continuous everywhere except at the break point $x = 0$, where we will check its continuity.

At $x = 0$.

L. H. L. $f(0 - 0) = \lim_{h \to 0} f(0 - h) = \lim_{h \to 0} (-h - 1) = -1.$

R. H. L. $f(0 + 0) = \lim_{h \to 0} f(0 + h) = \lim_{h \to 0} (h + 1) = 1.$

Since $f(0 - 0) \neq f(0 + 0)$, the function f is not continuous at $x = 0$.

Example 2. Show that the function f on the $[0, 1]$ defines as follows:

$$f(x) = 2rx, \quad \text{when} \quad \frac{1}{r + 1} \leq x < \frac{1}{r}, \ r = 1, 2, 3, \ldots,$$

$$f(0) = 0 \quad \text{and} \quad f(1) = 1,$$

is not continuous at $x = \frac{1}{2}, \frac{1}{3}, \frac{1}{4}, \ldots$.

Solution. Putting $r = 1, 2, 3, \ldots$, the given function is defined as follows:

$$f(x) = 2x, \quad \text{when} \quad \frac{1}{2} \leq x < 1,$$

$$= 4x, \quad \text{when} \quad \frac{1}{3} \leq x < \frac{1}{2},$$

$$= 6x, \quad \text{when} \quad \frac{1}{4} \leq x < \frac{1}{3},$$

$$\ldots\ldots\ldots\ldots\ldots\ldots\ldots$$

$$\ldots\ldots\ldots\ldots\ldots\ldots\ldots$$

Test the continuity at $x = \frac{1}{2}$.

$$\text{L. H. L. } f\left(\frac{1}{2} - 0\right) = \lim_{h \to 0} f\left(\frac{1}{2} - h\right)$$

$$= \lim_{h \to 0} 4\left(\frac{1}{2} - h\right) = 2.$$

$$\text{R. H. L. } f\left(\frac{1}{2} + 0\right) = \lim_{h \to 0} f\left(\frac{1}{2} + h\right)$$

$$= \lim_{h \to 0} 2\left(\frac{1}{2} + h\right) = 1.$$

Since $f(0 - 0) \neq f(0 + 0)$, the function f is not continuous at $x = \frac{1}{2}$.

To test the continuity at a general point $x = \frac{1}{r+1}$. We observe that

$$\text{L. H. L. } f\left(\frac{1}{r+1} - 0\right) = \lim_{h \to 0} f\left(\frac{1}{r+1} - h\right)$$

$$= \lim_{h \to 0} 2(r+1)\left(\frac{1}{r+1} - h\right) = 2.$$

$$\text{R. H. L. } f\left(\frac{1}{r+1} + 0\right) = \lim_{h \to 0} f\left(\frac{1}{r+1} + h\right)$$

$$= \lim_{h \to 0} 2r\left(\frac{1}{r+1} + h\right) = \frac{2r}{r+1}.$$

Since $f\left(\frac{1}{r+1} - 0\right) \neq f\left(\frac{1}{r+1} + 0\right)$, the function f is not continuous at $x = \frac{1}{r+1}$, where $r = 1, 2, 3, \ldots$.

So, the given function is discontinuous at $= \frac{1}{2}, \frac{1}{3}, \frac{1}{4}, \ldots$.

Example 3. Discuss the continuity of the function $f(x) = x[x]$, where $[x]$ denotes the greatest integer not greater than x.

Solution. By the definition of the greatest integer function, it is clear that it is continuous at every point except at the corner points $x = 0, \pm 1, \pm 2, \ldots$

Let $x = a$ be an integer. Then

L. H. L. $f(a - 0) = \lim_{h \to 0} f(a - h) = \lim_{h \to 0} (a - h)[a - h]$

$$= \lim_{h \to 0} (a - h) \lim_{h \to 0} [a - h] = a \times (a - 1) = a^2 - a.$$

R. H. L. $f(a + 0) = \lim_{h \to 0} f(a + h) = \lim_{h \to 0} (a + h)[a + h]$

$$= \lim_{h \to 0} (a + h) \lim_{h \to 0} [a + h] = a \times a = a^2.$$

So, the limit of the given function exists, if $a^2 - a = a^2 \Rightarrow a = 0$ and $f(0) = 0$.

Therefore, the given function is discontinuous at $x = \pm 1, \pm 2, \ldots$.

5.4 Algebra of Continuous Functions

Theorem 1. *Let f and g be two functions defined on the interval I. If f and g are continuous at $b \in I$, then their sum $f + g$ is also continuous at b.*

Proof. Since f is continuous at b, then the definition of continuity

$$\Rightarrow \lim_{x \to b} f(x) = f(b). \tag{5.6}$$

Since g is continuous at b, then the definition of continuity

$$\Rightarrow \lim_{x \to b} g(x) = g(b). \tag{5.7}$$

Now, we have

$$\lim_{x \to b} (f + g)(x) = \lim_{x \to b} (f(x) + g(x)) = \lim_{x \to b} f(x) + \lim_{x \to b} g(x)$$

$$= f(b) + g(b) = (f + g)(b),$$

by Eqs. (5.6) and (5.7).

$$\Rightarrow f + g \text{ is continuous at } b.$$

Theorem 2. *Let f and g be two functions defined on the interval I. If f and g are continuous at $b \in I$, then their difference $f - g$ is also continuous at b.*

Proof. Since the function f is continuous at b, then the definition of continuity

$$\Rightarrow \lim_{x \to b} f(x) = f(b). \tag{5.8}$$

Since g is continuous at b, then the definition of continuity

$$\Rightarrow \lim_{x \to b} g(x) = g(b). \tag{5.9}$$

Now, we have

$$\lim_{x \to b} (f - g)(x) = \lim_{x \to b} (f(x) - g(x)) = \lim_{x \to b} f(x) - \lim_{x \to b} g(x)$$
$$= f(b) - g(b) = (f - g)(b),$$

by Eqs. (5.8) and (5.9).

$$\Rightarrow f - g \text{ is continuous at } b.$$

Theorem 3. *Let f and g be two functions defined on the interval I. If f and g are continuous at $b \in I$, then their product fg is also continuous at b.*

Proof. Since f is continuous at b, then the definition of continuity

$$\Rightarrow \lim_{x \to b} f(x) = f(b). \tag{5.10}$$

Since g is continuous at b, then the definition of continuity

$$\Rightarrow \lim_{x \to b} g(x) = g(b). \tag{5.11}$$

Now, we have

$$\lim_{x \to b} (fg)(x) = \lim_{x \to b} (f(x)g(x)) = \lim_{x \to b} f(x) \lim_{x \to b} g(x)$$
$$= f(b)g(b) = (fg)(b),$$

by Eqs. (5.10) and (5.11).

$$\Rightarrow fg \text{ is continuous at } b.$$

Theorem 4. *Let f and g be two functions defined on the interval I and $g(b) \neq 0$. If f and g are continuous at $b \in I$, then $\frac{f}{g}$ is also continuous at b.*

Proof. Since f is continuous at b, then the definition of continuity

$$\Rightarrow \lim_{x \to b} f(x) = f(b). \tag{5.12}$$

Since g is continuous at b, then the definition of continuity

$$\Rightarrow \lim_{x \to b} g(x) = g(b), \quad \text{and} \quad g(b) \neq 0. \tag{5.13}$$

Now, we have

$$\lim_{x \to b} \left(\frac{f}{g} \right)(x) = \lim_{x \to b} \left(\frac{f(x)}{g(x)} \right) = \frac{\lim_{x \to b} (f(x))}{\lim_{x \to b} (g(x))}$$

$$= \frac{f(b)}{g(b)} = \left(\frac{f}{g} \right)(b),$$

by Eqs. (5.12) and (5.13).

$$\Rightarrow \frac{f}{g} \text{ is continuous at } b.$$

Theorem 5. *If f is continuous at a point b, then $|f|$ is also continuous at b, but the converse is not true.*

Proof. Let f be continuous at a point b. Then we will prove that $|f|$ is also continuous at b. If a and b are two real numbers, then we have the following relation:

$$||a| - |b|| \leq |a - b|. \tag{5.14}$$

Since f is continuous at the point b, therefore, for $\varepsilon > 0$, there exists a $\delta > 0$ such that

$$|f(x) - f(b)| < \varepsilon \quad \text{whenever} \quad |x - b| < \delta. \tag{5.15}$$

From Eqs. (5.14) and (5.15), we can write

$$||f(x)| - |f(b)|| \leq |f(x) - f(b)| < \varepsilon \text{ whenever } |x - b| < \delta.$$

$$\Rightarrow ||f(x)| - |f(b)|| < \varepsilon \text{ whenever } |x - b| < \delta.$$

$$\Rightarrow |f| \text{ is continuous at } x = b.$$

To show that the converse is not true, we consider the function f on **R** defined by

$$f(x) = \begin{cases} -2, & x < 1 \\ 2, & x \geq 1. \end{cases}$$

Therefore, $|f|(x) = |f(x)| = 2 \quad (\forall\, x \in \mathbf{R})$.

Then $\lim_{x \to 1} |f|(x) = \lim_{x \to 1}(2) = 2 = |f|(1)$.

$$\Rightarrow |f| \text{ is continuous at } x = 1.$$

Now, we have

$$\text{L. H. L. } f(1-0) = \lim_{h \to 0} f(1-h) = \lim_{h \to 0}(-2) = -2$$

and

$$\text{R. H. L. } f(1+0) = \lim_{h \to 0} f(1+h) = \lim_{h \to 0}(2) = 2.$$

Since $f(1-0) \neq f(1+0)$, the function f is not continuous at $x = 1$.

Thus, $|f|$ is continuous at $x = 1$, but f is not continuous at $x = 1$.

5.5 Sequential Continuity

Theorem 1. *A function f defined on an interval I is continuous at a point $b \in I$ if and only if each sequence $\langle b_n \rangle$ in I converges to b. Furthermore, the sequence $\langle f(b_n) \rangle$ converges to $f(b)$.*

Proof. Let f be a continuous function at a point $b \in I$ and let $\langle b_n \rangle$ be an arbitrary sequence in I that converges to b. Then we will show that the sequence $\langle f(b_n) \rangle$ converges to $f(b)$.

Since f is a continuous function at the point $b \in I$, therefore, for each $\varepsilon > 0$, there exists a $\delta > 0$ such that

$$|f(x) - f(b)| < \varepsilon \quad \text{whenever} \quad |x - b| < \delta. \tag{5.16}$$

Since $< b_n >$ is a sequence in I, by taking $x = b_n$ in Eq. (5.16), we get

$$|f(b_n) - f(b)| < \varepsilon \quad \text{whenever} \quad |b_n - b| < \delta. \tag{5.17}$$

Since $\langle b_n \rangle$ is a sequence in I that converges to b, therefore, for each $\delta > 0$, there exists a natural number m such that

$$|b_n - b| < \delta \quad \text{whenever} \quad n \geq m. \tag{5.18}$$

From Eqs. (5.17) and (5.18), we can write

$$|f(b_n) - f(b)| < \varepsilon \quad \text{whenever } n \geq m.$$
$$\Rightarrow \text{Sequence } \langle f(b_n) \rangle \text{ converges to } f(b).$$

To show the converse part, let the sequence $\langle f(b_n) \rangle$ converge to $f(b)$ whenever $\langle b_n \rangle$ converges to b. Then we will prove that f is continuous at $x = b$.

If possible, let *the function* f be not continuous at $x = b$. Then there is an $\varepsilon > 0$ such that, for every $\delta > 0$, there is an $x \in I$ such that

$$|f(x) - f(b)| \geq \varepsilon \quad \text{whenever } |x - b| < \delta. \tag{5.19}$$

Taking $\delta = \frac{1}{n}$, we find that, for each natural number n, there is a $b_n \in I$ such that

$$|f(b_n) - f(b)| \geq \varepsilon \quad \text{whenever } |b_n - b| < \frac{1}{n}. \tag{5.20}$$

Equation (5.20) shows that the sequence $\langle b_n \rangle$ converges to b, but the sequence $\langle f(b_n) \rangle$ does not converge to $f(b)$, which is a contradiction. Therefore, f is continuous at $x = b$.

Example 1 (Dirichlet's Function). Consider the function f on **R** defined by

$$f(x) = \begin{cases} 1, & x \text{ is rational} \\ -1, & x \text{ is irrational.} \end{cases}$$

Then show that f is not continuous everywhere in **R**. This function is called Dirichlet's function.

Solution. To show that f is not continuous everywhere in **R**, we consider the following two cases:

Case 1. Let $b \in$ **R** be an irrational number. We know that the set of rational numbers is dense in the set of real numbers. Therefore, every neighborhood of b contains infinite number of rational numbers.

That is, for each $n \in \mathbf{N}$, there exists a rational number b_n such that $b_n \in \left(b - \frac{1}{n}, b + \frac{1}{n}\right)$.

$$. \Rightarrow |b_n - b| < \frac{1}{n} \quad (\forall\, n \in \mathbf{N}).$$

$$\Rightarrow \lim_{n \to \infty} b_n = b.$$

That is, corresponding to each irrational number, we can always construct a sequence of rational numbers which converges to that number.

So, the sequence $\langle b_n \rangle$ of rational numbers converges to the irrational number b.

Now, $f(b_n) = 1$ $(\forall\, n \in \mathbf{N})$, because each b_n is a rational number.

$$\Rightarrow \lim_{n \to \infty} f(b_n) = 1.$$

But $f(b) = -1$, because b, is an irrational number.

$$\Rightarrow \lim_{n \to \infty} f(b_n) \neq f(b).$$

So f is not continuous at $x = b$ when b is an irrational number.

Case 2. Let $b \in \mathbf{R}$ be a rational number. We know that the set of irrational numbers is dense in the set of real numbers. Therefore, every neighborhood of b contains infinite number of irrational numbers. That is, for each $n \in \mathbf{N}$, there exists an irrational number b_n such that $b_n \in \left(b - \frac{1}{n}, b + \frac{1}{n}\right)$.

$$. \Rightarrow |b_n - b| < \frac{1}{n} \quad (\forall\, n \in \mathbf{N}).$$

$$\Rightarrow \lim_{n \to \infty} b_n = b.$$

That is, corresponding to each rational number, we can always construct a sequence of irrational numbers, which converges to that number.

So, the sequence $\langle b_n \rangle$ of irrational numbers converges to the rational number b.

Now, $f(b_n) = -1$ ($\forall\, n \in \mathbf{N}$), because each b_n is an irrational number.

$$\Rightarrow \lim_{n \to \infty} f(b_n) = -1.$$

But $f(b) = 1$, because b is a rational number.

$$\Rightarrow \lim_{n \to \infty} f(b_n) \neq f(b).$$

So, f is not continuous at $x = b$ when b is a rational number.

From Cases 1 and 2, it is clear that *the function f is not continuous everywhere in* $\mathbf{R}$.

Example 2. Consider the function f on $\mathbf{R}$ defined by

$$f(x) = \begin{cases} x, & x \text{ is rational} \\ 1, & x \text{ is irrational.} \end{cases}$$

Then show that f is not continuous in $\mathbf{R}$ except at $x = 1$.

Solution. We consider the following two cases:

Case 1. Let $b \neq 1 \in \mathbf{R}$ be an irrational number. We know that the set of rational numbers is dense in the set of real numbers. Therefore, every neighborhood of b contains infinite number of rational numbers. That is, for each $n \in \mathbf{N}$, there exists a rational number b_n such that $b_n \in \left(b - \frac{1}{n},\, b + \frac{1}{n} \right)$.

$$. \Rightarrow |b_n - b| < \frac{1}{n} \quad (\forall\, n \in \mathbf{N}).$$

$$\Rightarrow \lim_{n \to \infty} b_n = b.$$

So, the sequence $\langle b_n \rangle$ of rational numbers converges to the irrational number b.

Now, $f(b_n) = b_n$ ($\forall\, n \in \mathbf{N}$), because each b_n is a rational number.

$$\Rightarrow \lim_{n \to \infty} f(b_n) = \lim_{n \to \infty} b_n = b.$$

But $f(b) = 1$, because b is an irrational number.

$$\Rightarrow \lim_{n \to \infty} f(b_n) \neq f(b) \text{ because } b \neq 1.$$

So, f is not continuous at $x = b$ when b is an irrational number.

Case 2. Let $b \neq 1 \in \mathbf{R}$ be a rational number. We know that the set of irrational numbers is dense in the set of real numbers. Therefore, every neighborhood of b contains infinite number of irrational numbers. That is, for each $n \in \mathbf{N}$, there exists an irrational number b_n such that $b_n \in \left(b - \frac{1}{n}, b + \frac{1}{n}\right)$.

$$. \Rightarrow |b_n - b| < \frac{1}{n} \quad (\forall\, n \in \mathbf{N}).$$

$$\Rightarrow \lim_{n \to \infty} b_n = b.$$

So, the sequence $\langle b_n \rangle$ of irrational numbers converges to the rational number b.

Now, $f(b_n) = 1 \ (\forall\, n \in \mathbf{N})$, because each b_n is an irrational number.

$$\Rightarrow \lim_{n \to \infty} f(b_n) = 1.$$

But $f(b) = b$, because b is a rational number.

$$\Rightarrow \lim_{n \to \infty} f(b_n) \neq f(b), \text{ because } b \neq 1.$$

So, f is not continuous at $x = b$ when b is a rational number.

From Cases 1 and 2, it is clear that *the function f is not continuous at $b \neq 1 \in \mathbf{R}$.*

We now show that f is continuous at $x = 1$.

We have $f(1) = 1$. Let $\varepsilon > 0$ be any number and let $\delta = \varepsilon$. Then

$$|x - 1| < \delta \Rightarrow |f(x) - f(1)| = |x - 1| < \delta = \varepsilon,$$

where x is a rational number.

$$|x - 1| < \delta \Rightarrow |f(x) - f(1)| = |1 - 1| = 0 < \delta = \varepsilon,$$

where x is an irrational number.

Thus, for each $\varepsilon > 0$, there is a $\delta = \varepsilon > 0$ such that

$$|f(x) - f(1)| < \varepsilon \quad \text{whenever} \quad |x - 1| < \delta.$$

Therefore, the function f is continuous at $x = 1$.

Example 3. Consider the function f on $\mathbf{R}$ defined by

$$f(x) = \begin{cases} x, & x \text{ is rational} \\ 2x, & x \text{ is irrational.} \end{cases}$$

Then show that f is not continuous in $\mathbf{R}$ except at $x = 0$.

Solution. We consider the following two cases.

Case 1. Let $b \neq 0 \in \mathbf{R}$ be an irrational number. We know that the set of rational numbers is dense in the set of real numbers. Therefore, every neighborhood of b contains infinite number of rational numbers. That is, for each $n \in \mathbf{N}$, there exists a rational number b_n such that $b_n \in \left(b - \frac{1}{n}, \, b + \frac{1}{n} \right)$

$$. \Rightarrow |b_n - b| < \frac{1}{n} \quad (\forall \, n \in \mathbf{N}).$$

$$\Rightarrow \lim_{n \to \infty} b_n = b.$$

So, the sequence $\langle b_n \rangle$ of rational numbers converges to the irrational number b.

Now, $f(b_n) = b_n (\forall \, n \in \mathbf{N})$, because each b_n is a rational number.

$$\Rightarrow \lim_{n \to \infty} f(b_n) = \lim_{n \to \infty} b_n = b.$$

But $f(b) = 2b$, because b is an irrational number.

$$\Rightarrow \lim_{n \to \infty} f(b_n) \neq f(b), \text{ because } b \neq 2b \quad \text{if } b \neq 0.$$

So, f is not continuous at $x = b$ when b is an irrational number.

Case 2. Let $b \neq 0 \in \mathbf{R}$ be a rational number. We know that the set of irrational numbers is dense in the set of real numbers. Therefore, every neighborhood of b contains infinite number of irrational numbers. That is, for each $n \in \mathbf{N}$, there exists an irrational number b_n such that $b_n \in \left(b - \frac{1}{n}, \, b + \frac{1}{n} \right)$.

$$. \Rightarrow |b_n - b| < \frac{1}{n} \quad (\forall \, n \in \mathbf{N}).$$

$$\Rightarrow \lim_{n \to \infty} b_n = b.$$

So, the sequence $\langle b_n \rangle$ of irrational numbers converges to the rational number b.

Now, $f(b_n) = 2b_n$ $(\forall \, n \in \mathbf{N})$, because each b_n is an irrational number.

$$\Rightarrow \lim_{n \to \infty} f(b_n) = \lim_{n \to \infty} 2b_n = 2b.$$

But $f(b) = b$, because b is a rational number.

$$\Rightarrow \lim_{n \to \infty} f(b_n) \neq f(b), \text{ because } 2b \neq b \text{ if } b \neq 0.$$

So, f is not continuous at $x = b$ when b is a rational number.

From Cases 1 and 2, it is clear that f is not continuous at $b \neq 0 \in \mathbf{R}$.

We now show that the function f is continuous at $x = 0$.

We have $f(0) = 0$. Let $\varepsilon > 0$ be any number and *let* $\delta = \frac{\varepsilon}{2}$. Then

$$|x - 0| < \delta \Rightarrow |f(x) - f(0)| = |x - 0| < \delta = \frac{\varepsilon}{2} < \varepsilon,$$

where x is an rational number.

$$|x - 0| < \delta \Rightarrow |f(x) - f(0)| = |2x - 0| = |2x| < 2\delta = 2.\frac{\varepsilon}{2} = \varepsilon,$$

where x is an irrational number.

Thus, for each $\varepsilon > 0$, there is a $\delta = \frac{\varepsilon}{2} > 0$ such that

$$|f(x) - f(0)| < \varepsilon \text{ whenever } |x - 0| < \delta.$$

Therefore, the function f is continuous at $x = 0$.

5.6   Continuous Function on Closed Intervals

Theorem 1. *If a function is continuous on a closed interval, then it is bounded in that interval.*

Proof.   Let f be a function which is continuous on the closed interval $[a, b]$. If possible, let the function f be not bounded above. Then, for each $n \in \mathbf{N}$, we can find an element $x_n \in [a, b]$ such that

$$f(x_n) > n.$$

Now, $\langle x_n \rangle$ is a sequence in $[a, b]$, therefore, it is bounded. That is, $a \leq x_n \leq b \ (\forall\, n \in \mathbf{N})$.

Since every bounded sequence has a convergent subsequence, there exists a subsequence $\langle x_{n_k} \rangle$ of $\langle x_n \rangle$, which is convergent. Let $\langle x_{n_k} \rangle$ converge to $c \in [a, b]$.

Since $f(x_{n_k}) > n_k \, (\forall\, n \in \mathbf{N})$, therefore, the sequence $\langle f(x_{n_k}) \rangle$ diverges to $+\infty$.

So, we have a sequence $\langle x_n \rangle$ in $[a, b]$, which converges to $c \in [a, b]$. But the sequence $\langle f(x_{n_k}) \rangle$ does not converge to $f(c)$. This contradicts the fact that f is continuous in $[a, b]$. Therefore, our assumption is false.

So, the function f is bounded on $[a, b]$.

Theorem 2. *If a function is continuous on a closed interval, then it attains its bounds at least once in that interval.*

Proof. Let f be a function which is continuous on the closed interval $[a, b]$. If f is a constant function, then f attains its bounds at every point. Let us consider that f is not a constant function. Since f is continuous on the closed interval $[a, b]$, it is bounded therein by Theorem 1 above. Let k and K be its infimum and supremum, respectively. Then we will show that there exist some $p, q \in [a, b]$ such that

$$f(p) = k \quad \text{and} \quad f(q) = K.$$

Suppose, if possible, f does not attain its supremum. Then $f(x) \neq K$ ($\forall\, x \in [a, b]$).

Since $f(x) - K \neq 0$, consider the function $g(x) = \frac{1}{K - f(x)}$ ($\forall\, x \in [a, b]$).

The function g is positive and continuous on $[a, b]$. Therefore, g is bounded. Let $M > 0$ be supremum of the function g on $[a, b]$. That is,

$$g(x) < M \quad (\forall\, x \in [a, b]).$$

$$\Rightarrow \frac{1}{K - f(x)} \leq M \quad (\forall\, x \in [a, b]).$$

$$\Rightarrow f(x) \leq K - \frac{1}{M} \quad (\forall\, x \in [a, b]).$$

From above, we find that $K - \frac{1}{M}$ is the supremum of f on $[a, b]$, which contradicts the fact that K is the supremum of f on $[a, b]$. So, our assumption that f does not attain its supremum is false. That is, there exists some $q \in [a, b]$ such that

$$f(q) = K.$$

Similarly, we can show that there exists some $p \in [a, b]$ such that

$$f(p) = k.$$

Hence, the function f attains its bounds at least once in the closed interval $[a, b]$.

Theorem 3. *If a function* f *is continuous at some interior point* c, *of the closed interval* $[a, b]$ *and* $f(c) \neq 0$, *then there exists a neighborhood of* c *in which* f *has the same sign as* $f(c)$.

Proof. Since f is continuous at an interior point c, for each $\varepsilon > 0$ there exists a $\delta > 0$ such that

$$|f(x) - f(c)| < \varepsilon \quad (\forall \ |x - c| < \delta).$$
$$\Rightarrow \ f(c) - \varepsilon < f(x) < f(c) + \varepsilon$$

whenever $c - \delta < x < c + \delta$. \hfill (5.21)

Since $f(c) \neq 0$, there are two possibilities.

First, when $f(c) > 0$, we take $\varepsilon < f(c)$. Then, from the left inequality of Eq. (5.21), we have

$$0 < f(c) - \varepsilon < f(x) \quad \text{whenever } c - \delta < x < c + \delta.$$
$$\Rightarrow f(x) > 0 \quad \text{whenever } c - \delta < x < c + \delta.$$

In the second case when $f(c) < 0$, we take $\varepsilon < -f(c)$. Then, from the right inequality of Eq. (5.21), we have

$$f(x) < f(c) + \varepsilon < 0 \quad \text{whenever } \forall c - \delta < x < c + \delta.$$
$$\Rightarrow f(x) < 0 \quad \text{whenever } \forall c - \delta < x < c + \delta.$$

Theorem 4 (Intermediate value theorem). *If a function* f *is continuous on the closed interval* $[a, b]$ *and* $f(a) \neq f(b)$, *then it takes on every value between* $f(a)$ *and* $f(b)$.

Proof. Let M be any number between $f(a)$ and $f(b)$. That is, $f(a) < M < f(b)$ or $f(a) > M > f(b)$. Then we will show that there exists a number $c \in (a, b)$ such that $f(c) = M$.

Consider a function $g(x) = f(x) - M$.

Then g is continuous on $[a, b]$.

Also $g(a) = f(a) - M$ and $g(b) = f(b) - M$.

If $f(a) < M < f(b)$, then $g(a) < 0$ and $g(b) > 0$.

If $f(a) > M > f(b)$, then $g(a) > 0$ and $g(b) < 0$.

Since g is continuous and $g(a)$ and $g(b)$ are opposite in sign, there exists some $c \in (a, b)$ such that

$$g(c) = 0.$$
$$\Rightarrow f(c) - M = 0.$$
$$\Rightarrow f(c) = M \quad \text{for some } c \in (a, b).$$

Theorem 5 (Fixed point theorem). *If function f is continuous on the closed interval $[a, b]$ and $f(x) \in [a, b]\,(\forall\, x \in [a, b])$, then f has a fixed point, that is, there exists a number $\alpha \in [a, b]$ such that $f(\alpha) = \alpha$.*

Proof.   Let f be continuous on the closed interval $[a, b]$ and suppose that $f(x) \in [a, b]\,(\forall\, x \in [a, b])$. If $f(a) = a$ or $f(b) = b$, then there is nothing to prove, because either a or b will be a fixed point and we take $\alpha = a$ or b. Let $f(a) \neq a$ and $f(b) \neq b$. Then we will show that there exists some $\alpha \in (a, b)$ such that $f(\alpha) = \alpha$.

Let $f(a) > a$ and $f(b) < b$.

Consider $g(x) = f(x) - x$. Then g is continuous, because it is the difference of two continuous functions.

$$g(a) = f(a) - a > 0,$$
$$g(b) = f(b) - b < 0.$$

From the above two equations, it clear that 0 is an intermediate value of g on $[a, b]$.

Let g be a continuous function on $[a, b]$ and *suppose that* 0 is an intermediate value of g on $[a, b]$. Then, by the intermediate value theorem, there exists some $\alpha \in (a, b)$ such that

$$g(\alpha) = 0.$$
$$\Rightarrow f(\alpha) = \alpha.$$

Exercises

1. Show that $\lim_{x \to 1}(3x + 1) = 4$.
2. Evaluate $\lim_{x \to 0} \frac{1}{x}$.
   Ans. Does not exist.
3. Evaluate $\lim_{x \to 0} \frac{e^{\frac{1}{x}}}{e^{\frac{1}{x}} + 1}$.
   Ans. Does not exist.
4. Evaluate $\lim_{x \to 0}[x]$, where $[x]$ denotes the greatest integer not greater than x.
   Ans. Does not exist, because the limit from the left is -1 and the limit from the right is 0.
5. Show that the following function f is continuous at $x = 0$:

$$f(x) = \begin{cases} x \cos\left(\frac{1}{x}\right), & x \neq 0 \\ 0, & x = 0. \end{cases}$$

6. Show that the function $f(x) = |x| + |x + 1|$ is continuous everywhere.
7. Show that the following function f is continuous everywhere

$$f(x) = \begin{cases} -x, & x < -1 \\ 1, & -1 \leq x \leq 1 \\ x, & x > 1. \end{cases}$$

8. Find the values of the constants m and n for which the following function f is continuous everywhere:

$$f(x) = \begin{cases} 2x + 1, & x \leq 1 \\ mx^2 + n, & 1 < x < 3 \\ 5x + 2m, & x \geq 3. \end{cases}$$

   Ans. $m = 2$ and $n = 1$.
9. Show that the following function f is continuous at $x = b$:

$$f(x) = \begin{cases} \left(\frac{x^2}{b}\right) - b, & x \leq b \\ b - \left(\frac{b^2}{x}\right), & x > b. \end{cases}$$

10. Show that the following function f has a discontinuity of the first kind at $x = \frac{1}{2}$

$$f(x) = \begin{cases} x, & 0 \leq x < \dfrac{1}{2} \\ 1, & x = \dfrac{1}{2} \\ 1 - x, & \dfrac{1}{2} < x \leq 1. \end{cases}$$

11. Consider the function f on $\mathbf{R}$ defined by

$$f(x) = \begin{cases} x, & x \text{ is rational} \\ 1 - x, & x \text{ is irrational.} \end{cases}$$

Then show that the function f is not continuous in $\mathbf{R}$ except at $x = \frac{1}{2}$.

12. Consider the function f on $\mathbf{R}$ defined by

$$f(x) = \begin{cases} 1, & x \text{ is rational} \\ 0, & x \text{ is irrational.} \end{cases}$$

Then show that the function f is discontinuous at every point in $\mathbf{R}$.

Chapter 6

Uniform Continuity of Real Functions

In this chapter, we define the notion of uniform continuity of real functions. The continuity of a function is local in character, because continuity of a function at a point is meaningful. The uniform continuity is global in character, because we can talk about uniform continuity only over its domain. In this chapter, first we discuss uniform continuity. Then we give some examples of uniform and non-uniform continuous functions. Finally, we show that, on a closed interval. Every continuous function is also uniform continuous.

6.1 Uniform Continuity

Let f be a function defined on an interval I. Then the function f is said to be uniformly continuous on I if, for each $\varepsilon > 0$, there corresponds a $\delta > 0$ such that

$$|f(x) - f(y)| < \varepsilon \quad \text{whenever} \quad |x - y| < \delta \quad (\forall\, x, y \in I). \qquad (6.1)$$

In the definition of continuity of a function at a point c, the value of δ depends on both ε and the point c, whereas in the definition of uniform continuity of a function, the value of δ depends only on ε, and not on the pair of points in the given interval.

In continuity of a function on an interval I, we can find a δ corresponding to each point in I. It may seem that the infimum of the set consisting of the δ's corresponding to different points of I would work for the whole of the interval I. But the infimum may be zero. In general, therefore, a δ may not exist which can work for the entire interval. So, every continuous function may not be uniformly continuous.

Remark. To show that a function f is not uniformly continuous on an interval, we will show that, for some $\varepsilon > 0$, we cannot find a $\delta > 0$ satisfying Eq. (6.1). That is, for any $\delta > 0$, there exist $x, y \in I$ such that

$$|f(x) - f(y)| \geq \varepsilon \quad \text{whenever} \quad |x - y| < \delta.$$

Theorem. *If a function f is uniformly continuous on an interval I, then it is continuous there, but the converse is not true.*

Proof. Let a function f be uniformly continuous on an interval I. Then, for each $\varepsilon > 0$, there corresponds a $\delta > 0$ such that

$$|f(x) - f(y)| < \varepsilon \quad \text{whenever} \quad |x - y| < \delta. \qquad (6.2)$$

Let $b \in I$ be an arbitrary point. Taking $y = b$ in Eq. (6.2), we get

$$|f(x) - f(b)| < \varepsilon \quad \text{whenever} \quad |x - b| < \delta. \qquad (6.3)$$

Therefore, the function f is continuous at the point $b \in I$. Since b is an arbitrary point, so f is continuous on I.

To show that the converse is not true, consider a function $f : (0, 1] \to \mathbf{R}$ defined as follows:

$$f(x) = \frac{1}{x} \quad \text{for all } x \in (0, 1].$$

Let $c \in (0, 1]$ be an arbitrary point. Then

$$\lim_{x \to c} f(x) = \lim_{x \to c} \frac{1}{x} = \frac{1}{c} = f(c).$$

So, the function f is continuous at a point $c \in (0, 1]$. Since c is an arbitrary point, therefore, f is continuous on $(0, 1]$.

Let $\varepsilon = \frac{1}{3} > 0$ and δ be a positive number such that $n > \frac{1}{\delta}$ (which, by the Archimedean property, we can always find). Then

$$\left| \frac{1}{n} - \frac{1}{n+2} \right| = \left| \frac{2}{n(n+2)} \right| = \frac{2}{n(n+2)} < \frac{1}{n} < \delta.$$

Therefore, taking $x = \frac{1}{n}$ and $y = \frac{1}{n+2}$ as any two points of the interval $(0, 1]$, we get

$$|f(x) - f(y)| = \left| f\left(\frac{1}{n} \right) - f\left(\frac{1}{n+2} \right) \right| = |n - (n+2)| = 2 > \varepsilon.$$

$$\Rightarrow |f(x) - f(y)| > \varepsilon \quad \text{whenever } |x - y| < \delta.$$

Therefore, the function f is not uniformly continuous on $(0, 1]$.

Example 1. Show that the function $f(x) = \frac{1}{x^2}$ is uniformly continuous on $[b, \infty)$, where $b > 0$, but not uniformly continuous on $(0, \infty)$.

Solution. Let $x, y \in [b, \infty)$. Then $x, y \geq b > 0$.

$$\Rightarrow \frac{1}{x} \quad \text{and} \quad \frac{1}{y} \leq \frac{1}{b}.$$

$$\Rightarrow \frac{1}{xy} \leq \frac{1}{b^2} \quad \text{and} \quad \frac{1}{x} + \frac{1}{y} \leq \frac{2}{b}.$$

Now, we have

$$|f(x) - f(y)| = \left| \frac{1}{x^2} - \frac{1}{y^2} \right| = \left| \frac{1}{x} - \frac{1}{y} \right| \left| \frac{1}{x} + \frac{1}{y} \right|$$

$$\leq \frac{2}{b} \left| \frac{y - x}{xy} \right| \leq \frac{2}{b^3} |x - y|.$$

$$|f(x) - f(y)| \leq \frac{2}{b^3} |x - y|.$$

Let $\varepsilon > 0$ be given and let $\delta = \frac{\varepsilon b^3}{2}$. Then, whenever $|x - y| < \delta$, we obtain

$$|f(x) - f(y)| \leq \frac{2}{b^3} |x - y| < \frac{2\delta}{b^3} < \varepsilon.$$

Therefore, we have

$$|f(x) - f(y)| < \varepsilon \quad \text{whenever } |x - y| < \delta \quad (\forall \ x, y \in [b, \infty)).$$

Hence, the function $f(x) = \frac{1}{x^2}$ is uniformly continuous on $[b, \infty)$, where $b > 0$.

To show that function f is not uniformly continuous on $(0, \infty)$, we let $\varepsilon = \frac{1}{3} > 0$ and δ be a positive number, and *let* n be a natural number such that $n > \frac{1}{\delta}$ (which, by the Archimedean property, we can always find). Then

$$\left| \frac{1}{\sqrt{n}} - \frac{1}{\sqrt{n+2}} \right| = \left| \frac{\sqrt{n+2} - \sqrt{n}}{\sqrt{n}\sqrt{n+2}} \right| = \frac{2}{\sqrt{n}\sqrt{n+2}(\sqrt{n+2} + \sqrt{n})}.$$

Since $\sqrt{n}\sqrt{n+2} > \sqrt{n}$ and $\sqrt{n+2} + \sqrt{n} > \sqrt{n} + \sqrt{n} = 2\sqrt{n}$

$$\Rightarrow \quad \frac{1}{\sqrt{n}\sqrt{n+2}} < \frac{1}{\sqrt{n}} \quad \text{and} \quad \frac{1}{\sqrt{n+2} + \sqrt{n}} < \frac{1}{2\sqrt{n}}.$$

Therefore, we have

$$\left| \frac{1}{\sqrt{n}} - \frac{1}{\sqrt{n+2}} \right| = \frac{2}{\sqrt{n}\sqrt{n+2}(\sqrt{n+2} + \sqrt{n})} < \frac{2}{\sqrt{n}.2\sqrt{n}} < \frac{1}{n} < \delta.$$

Taking $x = \frac{1}{\sqrt{n}}$ and $y = \frac{1}{\sqrt{n+2}}$, as any two points of the interval $(0, \infty)$, we find that

$$|f(x) - f(y)| = \left| f\left(\frac{1}{\sqrt{n}} \right) - f\left(\frac{1}{\sqrt{n+2}} \right) \right| = |n - (n+2)| = 2 > \varepsilon.$$

$$\Rightarrow \quad |f(x) - f(y)| > \varepsilon \quad \text{whenever } \ |x - y| < \delta.$$

Therefore, the function f is not uniformly continuous on $(0, \infty)$.

Example 2. Show that the function $f(x) = x^3$ is uniformly continuous on $[-1, 1]$.

Solution. Let $x, y \in [-1, 1]$. Then $|x| \leq 1$ and $|y| \leq 1$.

$$\Rightarrow \quad |f(x) - f(y)| = |x^3 - y^3| = |x - y||x^2 + y^2 + xy|.$$

Since $|x| \leq 1$ and $|y| \leq 1$, therefore, we have

$$|x^2 + y^2 + xy| \leq |x|^2 + |y|^2 + |xy| \leq 3.$$

Now, we have

$$|f(x) - f(y)| = |x - y||x^2 + y^2 + xy| \leq 3|x - y|.$$

Let $\varepsilon > 0$ be given and let $\delta = \frac{\varepsilon}{3}$. Then, whenever $|x - y| < \delta$, we obtain

$$|f(x) - f(y)| \leq 3|x - y| < 3\delta < \varepsilon.$$

Therefore, we have

$$|f(x) - f(y)| < \varepsilon \quad \text{whenever } |x - y| < \delta, \ (\forall \, x, y \in [-1, 1]).$$

Hence, the function $f(x) = x^3$ is uniformly continuous on $[-1, 1]$.

Example 3. Show that the function $f(x) = 2x^2 - 3x + 5$ is uniformly continuous on $[-2, 2]$.

Solution. Let $x, y \in [-2, 2]$. Then $|x| \leq 2$ and $|y| \leq 2$.

$$\Rightarrow |f(x) - f(y)| = |(2x^2 - 3x + 5) - (2y^2 - 3y + 5)| = |x - y||2x + 2y - 3|.$$

Since $|x| \leq 2$ and $|y| \leq 2$, therefore, we have

$$|2x + 2y - 3| \leq 2|x| + 2|y| + 3 \leq 11.$$

Now, we have

$$|f(x) - f(y)| = |x - y||2x + 2y - 3| \leq 11|x - y|.$$

Let $\varepsilon > 0$ be given and let $\delta = \frac{\varepsilon}{11}$. Then, whenever $|x - y| < \delta$, we obtain

$$|f(x) - f(y)| \leq 11|x - y| < 11\delta < \varepsilon.$$

Therefore, we have

$$|f(x) - f(y)| < \varepsilon \quad \text{whenever } |x - y| < \delta, \ (\forall \, x, y \in [-2, 2]).$$

Hence, the function $f(x) = 2x^2 - 3x + 5$ is uniformly continuous on $[-2, 2]$.

Example 4. Show that the function $f(x) = \frac{x}{1+x^2}$ is uniformly continuous on $[0, 1]$.

Solution. Let $x, y \in [0, 1]$. Then $0 \le |x| \le 1$ and $0 \le |y| \le 1$.

$$\Rightarrow |f(x) - f(y)| = \left| \left(\frac{x}{1 + x^2} \right) - \left(\frac{y}{1 + y^2} \right) \right|$$

$$= |x - y| \left| \frac{1 - xy}{(1 + x^2)(1 + y^2)} \right|.$$

Since $0 \le |x| \le 1$ and $0 \le |y| \le 1$, therefore, we get

$$|1 - xy| \le 1 \quad \text{and} \quad \left| \frac{1}{(1 + x^2)(1 + y^2)} \right| \le 1.$$

Now, we have

$$|f(x) - f(y)| = |x - y| \left| \frac{1 - xy}{(1 + x^2)(1 + y^2)} \right| \le |x - y|.$$

Let $\varepsilon > 0$ be given and let $\delta = \varepsilon$. Then, whenever $|x - y| < \delta$, we obtain

$$|f(x) - f(y)| \le |x - y| < \delta = \varepsilon.$$

Therefore, we have

$$|f(x) - f(y)| < \varepsilon \quad \text{whenever } |x - y| < \delta, \ (\forall \, x, y \in [0, 1]).$$

Hence, the function $f(x) = \frac{x}{1+x^2}$ is uniformly continuous on $[0, 1]$.

Example 5. Show that the function $f(x) = \cos x$ is uniformly continuous on $\mathbf{R}$.

Solution. Let $x, y \in \mathbf{R}$.

$$\Rightarrow |f(x) - f(y)| = |\cos x - \cos y| = \left| -2 \sin \left(\frac{x + y}{2} \right) \sin \left(\frac{x - y}{2} \right) \right|$$

$$= 2 \left| \sin \left(\frac{x + y}{2} \right) \right| \left| \sin \left(\frac{x - y}{2} \right) \right|$$

$$\le 2.1 \left| \frac{x - y}{2} \right|,$$

because $|\sin \theta| \le 1$ and $|\sin \theta| \le |\theta|$.

Now, we have

$$|f(x) - f(y)| \le |x - y|.$$

Let $\varepsilon > 0$ be given and let $\delta = \varepsilon$. Then, whenever $|x - y| < \delta$, we obtain

$$|f(x) - f(y)| \le |x - y| < \delta < \varepsilon.$$

Therefore, we have

$$|f(x) - f(y)| < \varepsilon \quad \text{whenever } |x - y| < \delta, \ (\forall \ x, y \in \mathbf{R}).$$

Hence, the function $f(x) = \cos x$ is uniformly continuous on $\mathbf{R}$.

Example 6. Show that the function $f(x) = \sin^2 x$ is uniformly continuous on $\mathbf{R}$.

Solution. Let $x, y \in \mathbf{R}$. Then

$$
\begin{aligned}
|f(x) - f(y)| &= |\sin^2 x - \sin^2 y| = |(\sin x - \sin y)(\sin x + \sin y)| \\
&= |(\sin x - \sin y)||(\sin x + \sin y)| \\
&\le 2 \left| \sin\left(\frac{x-y}{2}\right) \right| \left| \cos\left(\frac{x+y}{2}\right) \right| (|\sin x| + |\sin y|) \\
&\le 2 \left| \frac{x-y}{2} \right| .1.2,
\end{aligned}
$$

because $|\sin \theta| \le 1$, $|\sin \theta| \le |\theta|$ and $|\cos \theta| \le 1$.
   Now, we have

$$|f(x) - f(y)| \le 2|x - y|.$$

Let $\varepsilon > 0$ be given and let $\delta = \frac{\varepsilon}{2}$. Then, whenever $|x - y| < \delta$, we obtain

$$|f(x) - f(y)| \le 2|x - y| < 2\delta < \varepsilon.$$

Therefore, we have

$$|f(x) - f(y)| < \varepsilon \quad \text{whenever } |x - y| < \delta, \ (\forall \, x, y \in \mathbf{R}).$$

Hence, the function $f(x) = \sin^2 x$ is uniformly continuous on $\mathbf{R}$.

Example 7. Show that the function $f(x) = \sin x^2$ is not uniformly continuous on $[0, \infty)$.

Solution. To show that function f is not uniformly continuous on $[0, \infty)$, let $\varepsilon = \frac{1}{3} > 0$, δ be a positive number, and n be a natural number such that $n > \frac{\pi}{\delta^2}$. Then

$$\left| \sqrt{\frac{n\pi}{2}} - \sqrt{\frac{(n+1)\pi}{2}} \right| = \left| \frac{\frac{n\pi}{2} - \frac{(n+1)\pi}{2}}{\sqrt{\frac{n\pi}{2}} + \sqrt{\frac{(n+1)\pi}{2}}} \right| = \left| \frac{\frac{\pi}{2}}{\sqrt{\frac{n\pi}{2}} + \sqrt{\frac{(n+1)\pi}{2}}} \right|$$

$$\leq \frac{\frac{\pi}{2}}{\sqrt{\frac{n\pi}{2}} + \sqrt{\frac{n\pi}{2}}} = \frac{\frac{\pi}{2}}{2\sqrt{\frac{n\pi}{2}}} < \frac{\pi}{\sqrt{n\pi}}.$$

Therefore, we have

$$\left| \sqrt{\frac{n\pi}{2}} - \sqrt{\frac{(n+1)\pi}{2}} \right| < \sqrt{\frac{\pi}{n}} < \delta.$$

Taking $x = \sqrt{\frac{n\pi}{2}}$ and $y = \sqrt{\frac{(n+1)\pi}{2}}$ as any two points of the interval $[0, \infty)$, we see that

$$|f(x) - f(y)| = \left| f\left(\sqrt{\frac{n\pi}{2}}\right) - f\left(\sqrt{\frac{(n+1)\pi}{2}}\right) \right|$$

$$= \left| \sin \frac{n\pi}{2} - \sin \frac{(n+1)\pi}{2} \right| = 1 > \varepsilon.$$

$$\Rightarrow |f(x) - f(y)| > \varepsilon \quad \text{whenever } |x - y| < \delta.$$

Therefore, the function f is not uniformly continuous on $[0, \infty)$.

Example 8. Show that the function $f(x) = \sin \frac{1}{x}$ is not uniformly continuous on $(0, \infty)$.

Solution. To show that the function f is not uniformly continuous on $(0, \infty)$, let $\varepsilon = \frac{1}{3} > 0$, δ be a positive number, and n be a natural number such that $\delta > \frac{2}{n(n+1)\pi}$. Then we have

$$\left| \frac{2}{n\pi} - \frac{2}{(n+1)\pi} \right| = 2 \left| \frac{(n+1)\pi - n\pi}{(n+1)\pi n\pi} \right| = \frac{2}{n(n+1)\pi}.$$

Therefore, we obtain

$$\left| \frac{2}{n\pi} - \frac{2}{(n+1)\pi} \right| = \frac{2}{n(n+1)\pi} < \delta.$$

Taking $x = \frac{2}{n\pi}$ and $y = \frac{2}{(n+1)\pi}$ as any two points of the interval $(0, \infty)$, we have

$$|f(x) - f(y)| = \left| f\left(\frac{2}{n\pi}\right) - f\left(\frac{2}{(n+1)\pi}\right) \right|$$

$$= \left| \sin \frac{n\pi}{2} - \sin \frac{(n+1)\pi}{2} \right| = 1 > \varepsilon.$$

$$\Rightarrow \quad |f(x) - f(y)| > \varepsilon \quad \text{whenever } |x - y| < \delta.$$

Therefore, the function f is not uniformly continuous on $(0, \infty)$.

Example 9. Show that the function $f(x) = \sin \frac{1}{x^2}$ is not uniformly continuous on $(0, \infty)$.

Solution. To show that the function f is not uniformly continuous on $(0, \infty)$, let $\varepsilon = \frac{1}{3} > 0$, δ be a positive number, and n be a natural number such that $> \frac{1}{\sqrt{2\pi}\delta}$. Then we have

$$\left| \sqrt{\frac{2}{n\pi}} - \sqrt{\frac{2}{(n+1)\pi}} \right| = \sqrt{2} \left| \frac{\sqrt{(n+1)\pi} - \sqrt{n\pi}}{\sqrt{n\pi}\sqrt{(n+1)\pi}} \right|$$

$$= \frac{\sqrt{2}}{\sqrt{n}\sqrt{n+1}} \frac{1}{\left(\sqrt{n\pi} + \sqrt{(n+1)\pi}\right)}$$

$$= \sqrt{\frac{2}{\pi}} \frac{1}{\sqrt{n}\sqrt{n+1}} \frac{1}{\left(\sqrt{n} + \sqrt{(n+1)}\right)}.$$

Since $\sqrt{n}\sqrt{n+1} > \sqrt{n}$ and $\sqrt{n} + \sqrt{n+1} > 2\sqrt{n}$

$$\Rightarrow \quad \frac{1}{\sqrt{n}\sqrt{n+1}} < \frac{1}{\sqrt{n}} \quad \text{and} \quad \frac{1}{\sqrt{n} + \sqrt{n+1}} < \frac{1}{2\sqrt{n}}.$$

Therefore, we have

$$\left| \sqrt{\frac{2}{n\pi}} - \sqrt{\frac{2}{(n+1)\pi}} \right| = \sqrt{\frac{2}{\pi}} \frac{1}{\sqrt{n}\sqrt{n+1}} \frac{1}{\left(\sqrt{n} + \sqrt{(n+1)}\right)}$$

$$< \sqrt{\frac{2}{\pi}} \cdot \frac{1}{\sqrt{n}} \cdot \frac{1}{2\sqrt{n}} = \frac{1}{\sqrt{2\pi}n} < \delta.$$

Taking $x = \sqrt{\frac{2}{n\pi}}$ and $y = \sqrt{\frac{2}{(n+1)\pi}}$ as any two points of the interval $(0, \infty)$, then

$$|f(x) - f(y)| = \left| f\left(\sqrt{\frac{2}{n\pi}}\right) - f\left(\sqrt{\frac{2}{(n+1)\pi}}\right) \right|$$

$$= \left| \sin\frac{n\pi}{2} - \sin\frac{(n+1)\pi}{2} \right| = 1 > \varepsilon.$$

$$\Rightarrow \quad |f(x) - f(y)| > \varepsilon \quad \text{whenever} \quad |x - y| < \delta.$$

Therefore, the function f is not uniformly continuous on $(0, \infty)$.

Remark. In Theorem 1, we have shown that a uniformly continuous function is continuous, but the converse is not true. That is, a continuous function need not be uniformly continuous. Now, the question arises as to under what conditions a continuous function will be uniformly continuous. The answer to this question is given in the following section.

6.2 Uniform Continuity on a Closed Bounded Interval

In this section, we will see that continuity and uniform continuity are the same on a closed bounded interval. That is, on a closed bounded interval, a function is continuous if and only if it is uniformly continuous.

Theorem 1. *If a function f is continuous on the closed interval $[a, b]$, then it is uniformly continuous on $[a, b]$.*

Proof. Let the function f be continuous on the closed interval $[a, b]$. If possible, let f be not uniformly continuous on $[a, b]$. Then there exists an $\varepsilon > 0$ and there corresponds a $\delta > 0$ such that

$$|f(x) - f(y)| \geq \varepsilon \quad \text{whenever} \quad |x - y| < \delta. \tag{6.4}$$

In particular, for each natural number n, we can find $x_n, y_n \in [a, b]$ such that

$$|f(x_n) - f(y_n)| \geq \varepsilon \quad \text{whenever} \quad |x - y| < \frac{1}{n}. \tag{6.5}$$

Since $a \leq x_n, y_n \leq b$, therefore, the sequences $\langle x_n \rangle$ and $\langle y_n \rangle$ are bounded. By the Bolzano–Weierstrass property, the sequences $\langle x_n \rangle$

and $\langle y_n \rangle$ have a limit point. Let p and q be the limit points of the sequences $\langle x_n \rangle$ and $\langle y_n \rangle$, respectively. Since $[a, b]$ is a closed set, therefore, $p, q \in [a, b]$.

Since p is a limit point of sequence $\langle x_n \rangle$, therefore, there exists a convergent subsequence $\langle x_{n_l} \rangle$ converging to p. That is,

$$\lim_{l \to \infty} x_{n_l} = p. \tag{6.6}$$

Since q is a limit point of the sequence $\langle y_n \rangle$, therefore, there exists a convergent subsequence $\langle y_{n_l} \rangle$ converging to q. That is,

$$\lim_{l \to \infty} y_{n_l} = q. \tag{6.7}$$

Now, from Eq. (6.3), for all k we have

$$|f(x_{n_l}) - f(y_{n_l})| \geq \varepsilon \quad \text{whenever} \quad |x_{n_l} - y_{n_l}| < \frac{1}{n_l}. \tag{6.8}$$

From the second inequality in Eq. (6.8), we have

$$\lim_{l \to \infty} x_{n_l} = \lim_{l \to \infty} y_{n_l}. \tag{6.9}$$

From Eqs. (6.6), (6.7), and (6.9), we get

$$p = q.$$

From the first inequality in Eq. (6.8), we have

$$\lim_{l \to \infty} f(x_{n_l}) \neq \lim_{l \to \infty} f(y_{n_l}).$$

Therefore, we have two sequences $\langle x_{n_l} \rangle$ and $\langle y_{n_l} \rangle$, which converge to the same limit p. But the sequences $\langle f(x_{n_l}) \rangle$ and $\langle f(y_{n_l}) \rangle$ do not converge to the same limit. This contradicts the fact that f is continuous. Therefore, our assumption that f is not uniformly continuous is false. Hence, the function f is uniformly continuous on $[a, b]$.

Theorem 2. *A function f is continuous on a closed interval $[a, b]$ if and only if it is uniformly continuous on $[a, b]$.*

Proof. Refer to the theorems in Sections 6.1 and 6.2.

Exercises

1. Show that the function $f(x) = \frac{x}{x+1}$ is uniformly continuous for $x \in [0, 2]$.
2. Show that the function $f(x) = x^2$ is uniformly continuous on $[-1, 1]$.
3. Show that the function $f(x) = \sqrt{x}$ is uniformly continuous on $[1, 3]$.
4. Show that the function $f(x) = \sqrt{x}$ is uniformly continuous on $[0, \infty)$.

Chapter 7

Differentiability of the Real Functions

In this chapter, we define the notion of derivative of a real function. We are already familiar with the geometrical concept of derivative. Here, we mainly discuss the mathematical concept of the derivative. First, we introduce the definition and examples of derivatives. In continuation, we discuss the algebra of derivatives. Further, we define the meaning of the sign of derivatives. At the end of the chapter, read the most important theorems on derivatives, Darboux's theorem and intermediate value theorem.

7.1 Derivative at an Interior Point

Let $f : [a, b] \rightarrow \mathbf{R}$. Then a function f is said to be differentiable at an interior point c, that is, $a < c < b$, if

$$\lim_{x \to c} \frac{f(x) - f(c)}{x - c},$$

exists and the value of the limit is called derivative of the function f at point c. The derivative of f at point c is denoted by $f'(c)$.

Right-hand derivative. Let $f : [a, b] \rightarrow \mathbf{R}$. Then a function f is said to be differentiable from the right at an interior point c if

$$\lim_{x \to c+0} \frac{f(x) - f(c)}{x - c} \quad \text{or} \quad \lim_{h \to 0} \frac{f(c + h) - f(c)}{h}$$

exists. The value of this limit is called right-hand derivative of the function f at the point c. The right-hand derivative of f at the point c is denoted by $f'(c + 0)$.

Left-hand derivative. Let $f : [a, b] \to \mathbf{R}$. Then a function f is said to be differentiable from the left at an interior point c if

$$\lim_{x \to c - 0} \frac{f(x) - f(c)}{x - c} \quad \text{or} \quad \lim_{h \to 0} \frac{f(c) - f(c - h)}{h}$$

exists. The value of this limit is called the left-hand derivative of the function f at the point c. The left-hand derivative of f at the point c is denoted by $f'(c - 0)$.

We say that a function f is differentiable at the point c if the left-hand and the right-hand derivatives of the function at the point c are equal. That is,

$$f'(c - 0) = f'(c + 0).$$

Derivative at boundary points. Let $f : [a, b] \to \mathbf{R}$. Then a function f is said to be differentiable at the boundary point a if the right-hand derivative exists at the point a. That is,

$$\lim_{x \to a + 0} \frac{f(x) - f(a)}{x - a} \quad \text{or} \quad \lim_{h \to 0} \frac{f(a + h) - f(a)}{h}$$

exists. The value of this limit is called the derivative of the function f at the point a and it is denoted by $f'(a + 0)$.

The function f is said to be differentiable at the boundary point b if the left-hand derivative of the function f exists at the point b. That is,

$$\lim_{x \to b - 0} \frac{f(x) - f(b)}{x - b} \quad \text{or} \quad \frac{f(b) - f(b - h)}{h}$$

exists. The value of this limit is called the derivative of the function f at the point b, and it is denoted by $f'(b - 0)$.

Theorem 1. *Continuity is a necessary but not a sufficient condition for the differentiability of a function at a point.*

Proof. Let f be a differentiable function at a point c. Then we will show that f is also continuous at the point c. If f is differentiable at the point c then, by the definition of derivative, the limit

$$\lim_{x \to c} \frac{f(x) - f(c)}{x - c}$$

exists. Let

$$\lim_{x \to c} \frac{f(x) - f(c)}{x - c} = f'(c).$$

Now, we have

$$\lim_{x \to c} \{f(x) - f(c)\} = \lim_{x \to c} \left\{ \frac{f(x) - f(c)}{x - c}(x - c) \right\},$$

$$= \lim_{x \to c} \left\{ \frac{f(x) - f(c)}{x - c} \right\} \cdot \lim_{x \to c} (x - c),$$

$$= f'(c).0 = 0.$$

$$\Rightarrow \lim_{x \to c} \{f(x) - f(c)\} = 0.$$

$$\Rightarrow \lim_{x \to c} f(x) = f(c).$$

Therefore, f is continuous at point c.

To show that the converse is not true, consider the function $f(x) = |x - 1|, \forall x \in \mathbf{R}$ which is continuous at $x = 1$, but not differentiable at $x = 1$. Using the property of modulus, we can redefine the above function as follows:

$$f(x) = \begin{cases} 1 - x, & x < 1, \\ x - 1, & x \geq 1. \end{cases}$$

To check its continuity at $\boldsymbol{x = 1}$

L. H. L. $f(1 - 0) = \lim_{h \to 0} f(1 - h) = \lim_{h \to 0} (1 - (1 - h)) = \lim_{h \to 0} (h) = 0,$

R. H. L. $f(1 + 0) = \lim_{h \to 0} f(1 + h) = \lim_{h \to 0} ((1 + h) - 1) = \lim_{h \to 0} (h) = 0,$

and $f(1) = 0$.

Since $f(1 - 0) = f(1 + 0) = f(1) = 0$, therefore, the function f is continuous at $\boldsymbol{x = 1}$.

To check its differentiability at **x = 1**

$$\text{L. H. D. } f'(1-0) = \lim_{h \to 0} \frac{f(1) - f(1-h)}{h} = \lim_{h \to 0} \left(\frac{0-h}{h} \right)$$

$$= \lim_{h \to 0} (-1) = -1,$$

and

$$\text{R. H. D. } f'(1+0) = \lim_{h \to 0} \frac{f(1+h) - f(1)}{h} = \lim_{h \to 0} \left(\frac{h-0}{h} \right)$$

$$= \lim_{h \to 0} (1) = 1.$$

Since $f'(1-0) \neq f'(1+0)$, the function f is not differentiable at $x = 1$.

Therefore, $f(x) = |x-1| \ \forall \, x \in \mathbf{R}$ is continuous at $x = 1$ but not differentiable at $x = 1$.

So, continuity is a necessary but not a sufficient condition for the differentiability of a function at a point.

Example 1. Show that the function $f(x) = |x-1| + |x-2|$ is continuous, but not differentiable, at $x = 1$ and $x = 2$.

Solution. Using the property of modulus, we can redefine the above function as follows:

$$f(x) = \begin{cases} 3 - 2x, & x < 1, \\ 1, & 1 \leq x < 2 \\ 2x - 3, & x \geq 2. \end{cases}$$

First, we will check its continuity and differentiability at **x = 1**.

$$\text{L. H. L. } f(1-0) = \lim_{h \to 0} f(1-h) = \lim_{h \to 0} (3 - 2(1-h))$$

$$= \lim_{h \to 0} (1 + 2h) = 1.$$

$$\text{R. H. L. } f(1+0) = \lim_{h \to 0} f(1+h) = \lim_{h \to 0} (1) = 1,$$

and $f(1) = 1$.

Since $f(1-0) = f(1+0) = f(1) = 1$, the function f is continuous at $x = 1$. Now,

$$\text{L. H. D. } f'(1-0) = \lim_{h \to 0} \frac{f(1) - f(1-h)}{h}$$

$$= \lim_{h \to 0} \left(\frac{1 - (3 - 2(1 - h))}{h} \right)$$

$$= \lim_{h \to 0} \left(\frac{-2h}{h} \right) = \lim_{h \to 0} (-2) = -2,$$

and

$$\text{R. H. D. } f'(1+0) = \lim_{h \to 0} \frac{f(1+h) - f(1)}{h}$$

$$= \lim_{h \to 0} \left(\frac{1 - 1}{h} \right) = \lim_{h \to 0} \left(\frac{0}{h} \right) = \lim_{h \to 0} (0) = 0.$$

Since $f'(1 - 0) \neq f'(1 + 0)$, the function f is not differentiable at $x = 1$.

At x = 2

$$\text{L. H. L. } f(2 - 0) = \lim_{h \to 0} f(2 - h) = \lim_{h \to 0} (1) = 1.$$

$$\text{R. H. L. } f(2 + 0) = \lim_{h \to 0} f(2 + h) = \lim_{h \to 0} (2(2 + h) - 3)$$

$$= \lim_{h \to 0} (2h + 1) = 1,$$

and $f(2) = 2 \times 2 - 3 = 1$.

Since $f(2-0) = f(2+0) = f(2) = 1$, the function f is continuous at $x = 2$.

$$\text{L. H. D. } f'(2 - 0) = \lim_{h \to 0} \frac{f(2) - f(2 - h)}{h} = \lim_{h \to 0} \left(\frac{1 - 1}{h} \right)$$

$$= \lim_{h \to 0} \left(\frac{0}{h} \right) = \lim_{h \to 0} (0) = 0,$$

and

$$\text{R. H. D. } f'(2+0) = \lim_{h\to 0} \frac{f(2+h) - f(2)}{h}$$

$$= \lim_{h\to 0} \left(\frac{(2(2+h) - 3) - 1}{h} \right)$$

$$= \lim_{h\to 0} \left(\frac{2h}{h} \right) = \lim_{h\to 0} (2) = 2.$$

Since $f'(2-0) \neq f'(2+0)$, the function f is not differentiable at $x = 2$.

So, the given function is continuous, but not differentiable, at $x = 1$ and $x = 2$.

Example 2. Discuss the continuity and differentiability of the following function at $x = 2$ and 4.

$$f(x) = \begin{cases} 2x - 3, & 0 \leq x \leq 2 \\ x^2 - 3, & 2 < x \leq 4. \end{cases}$$

Solution. First, we will check its continuity and derivability at **x = 2**.

$$\text{L. H. L. } f(2-0) = \lim_{h\to 0} f(2-h) = \lim_{h\to 0} (2(2-h) - 3)$$

$$= \lim_{h\to 0} (1 - 2h) = 1.$$

$$\text{R. H. L. } f(2+0) = \lim_{h\to 0} f(2+h) = \lim_{h\to 0} ((2+h)^2 - 3)$$

$$= \lim_{h\to 0} (h^2 + 4h + 1) = 1,$$

and $f(2) = 4 - 3 = 1$.

Since $f(2-0) = f(2+0) = f(2) = 1$, the function f is continuous at $x = 1$.

Now,

$$\text{L. H. D. } f'(2-0) = \lim_{h \to 0} \frac{f(2) - f(2-h)}{h} = \lim_{h \to 0} \left(\frac{1 - (1 - 2h)}{h} \right)$$

$$= \lim_{h \to 0} \left(\frac{2h}{h} \right) = \lim_{h \to 0} (2) = 2,$$

and

$$\text{R. H. D. } f'(2+0) = \lim_{h \to 0} \frac{f(2+h) - f(2)}{h}$$

$$= \lim_{h \to 0} \left(\frac{\left(h^2 + 4h + 1 \right) - 1}{h} \right)$$

$$= \lim_{h \to 0} \left(\frac{h^2 + 4h}{h} \right) = \lim_{h \to 0} (h + 4) = 4.$$

Since $f'(2 - 0) \neq f'(2 + 0)$, the function f is not differentiable at $x = 2$.

Now, at $\boldsymbol{x = 4}$

$$\text{L. H. L. } f(4-0) = \lim_{h \to 0} f(4 - h) = \lim_{h \to 0} ((4 - h)^2 - 3)$$

$$= \lim_{h \to 0} (h^2 - 8h + 13) = 13,$$

and $f(4) = 16 - 3 = 13$.

Since $x = 4$ is a boundary point and $f(4 - 0) = f(4) = 13$, the function f is continuous at $x = 4$.

Now,

$$\text{L. H. D. } f'(4-0) = \lim_{h \to 0} \frac{f(4) - f(4-h)}{h}$$

$$= \lim_{h \to 0} \left(\frac{13 - \left(h^2 - 8h + 13 \right)}{h} \right)$$

$$= \lim_{h \to 0} \left(\frac{-h^2 + 8h}{h} \right) = \lim_{h \to 0} (-h + 8) = 8.$$

Since $x = 4$ is a boundary point, so we will calculate only its one-side derivative. Therefore, $f'(4) = 8$.

Example 3. Determine the value of n for which the following function f is continuous and differentiable at $x = 0$

$$f(x) = \begin{cases} x^n \sin\left(\dfrac{1}{x}\right), & x \neq 0, \\ 0, & x = 0. \end{cases}$$

Solution. Here, we have

$$\text{L. H. L. } f(0-0) = \lim_{h \to 0} f(0-h) = \lim_{h \to 0}\left((-h)^n \sin\left(\frac{1}{-h}\right)\right)$$

$$= (-1)^{n+1} \times \lim_{h \to 0}\left(h^n \sin\left(\frac{1}{h}\right)\right)$$

$$= (-1)^{n+1} \times \lim_{h \to 0}(h^n) \times \lim_{h \to 0}\left(\sin\left(\frac{1}{h}\right)\right),$$

$$\text{L. H. L. } f(0+0) = \lim_{h \to 0} f(0+h) = \lim_{h \to 0}\left((h)^n \sin\left(\frac{1}{h}\right)\right)$$

$$= (1)^{n+1} \times \lim_{h \to 0}\left(h^n \sin\left(\frac{1}{h}\right)\right)$$

$$= \lim_{h \to 0}(h^n) \times \lim_{h \to 0}\left(\sin\left(\frac{1}{h}\right)\right).$$

If f is continuous at $x = 0$, then $f(0-0) = f(0+0) = f(0) = 0$. That is, both limits will be zero.

$$\Rightarrow \lim_{h \to 0}(h^n) \times \lim_{h \to 0}\left(\sin\left(\frac{1}{h}\right)\right) = 0$$

$$\Rightarrow \lim_{h \to 0}(h^n) \times l = 0, \text{ because } \lim_{h \to 0} \sin\left(\frac{1}{h}\right) = l$$

and l is a finite number lying between -1 to 1.
   This above limit is zero if $n > 0$.
   Therefore, the function f is continuous at $x = 0$ if $n > 0$.

For differentiability, we see that

L. H. D. $f'(0-0) = \lim_{h \to 0} \dfrac{f(0) - f(0-h)}{h}$

$$= \lim_{h \to 0} \left(\frac{0 - (-h)^n \sin\left(\frac{1}{-h}\right)}{h} \right)$$

$$= (-1)^{n+2} \times \lim_{h \to 0} \left(h^{n-1} \sin\left(\frac{1}{h}\right) \right)$$

$$= (-1)^{n+2} \times \lim_{h \to 0} (h^{n-1}) \times \lim_{h \to 0} \left(\sin\left(\frac{1}{h}\right) \right)$$

and

R. H. D. $f'(0+0) = \lim_{h \to 0} \dfrac{f(0+h) - f(0)}{h} = \lim_{h \to 0} \left(\dfrac{(h)^n \sin\left(\frac{1}{h}\right) - 0}{h} \right)$

$$= \lim_{h \to 0} \left(h^{n-1} \sin\left(\frac{1}{h}\right) \right) = \lim_{h \to 0} (h^{n-1})$$

$$\times \lim_{h \to 0} \left(\sin\left(\frac{1}{h}\right) \right)$$

If f is differentiable at $x = 0$, then $f'(0-0) = f'(0+0)$. That is, both limits will be equal. This will be the case when $n - 1 > 0$, because $\lim_{h \to 0}(h^{n-1}) = 0$ for $n - 1 > 0$. That is, for $n > 1$, we have

$$f'(0-0) = f'(0+0) = 0.$$

Therefore, the function f is differentiable at $x = 0$ if $n > 1$.

Consequently, the function f is continuous at $x = 0$ if $n > 0$, and differentiable at $x = 0$ if $n > 1$.

Example 4. Find the value of the constants m and n for which the following function f is differentiable at $x = 0$.

$$f(x) = \begin{cases} e^x + m \sin x, & x < 0 \\ n(x-1)^2 + x - 2, & x \geq 0. \end{cases}$$

Solution. The given function f is differentiable at $x = 0$. Therefore, it must be continuous there. For the continuity at the origin, we have

$$\text{L. H. L. } f(0-0) = \lim_{h \to 0} f(0-h) = \lim_{h \to 0} (e^{-h} + m \sin(-h))$$

$$= 1 + m.0 = 1,$$

$$\text{R. H. L. } f(0+0) = \lim_{h \to 0} f(0+h) = \lim_{h \to 0} (n(h-1)^2 + h - 2)$$

$$= n.1 + 0 - 2 = n - 2,$$

and $f(0) = n - 2$.

The given function f is continuous at $x = 0$. Therefore, $f(0-0) = f(0+0) = f(0)$.

$$\Rightarrow n - 2 = 1$$

$$\Rightarrow n = 3.$$

Now, we have $f(0) = 3 - 2 = 1$.

Next, for the derivative at $x = 0$, we have

$$\text{L. H. D. } f'(0-0) = \lim_{h \to 0} \frac{f(0) - f(0-h)}{h} \quad \lim_{h \to 0} \left(\frac{1 - e^{-h} - m \sin(-h)}{h} \right)$$

$$= \lim_{h \to 0} \left(\frac{1 - e^{-h}}{h} \right) + \lim_{h \to 0} \left(\frac{m \sin(h)}{h} \right)$$

$$= 1 + m,$$

because $\lim_{h \to 0} \left(\frac{1-e^{-h}}{h} \right) = 1$ and $\lim_{h \to 0} \left(\frac{m \sin(h)}{h} \right) = m$,

and

$$\text{R. H. D. } f'(0+0) = \lim_{h \to 0} \frac{f(0+h) - f(0)}{h}$$

$$= \lim_{h \to 0} \left(\frac{(n(h-1)^2 + h - 2) - 1}{h} \right) = -5.$$

The given function f is differentiable at $x = 0$. Therefore, $f'(0-0) = f'(0+0)$.

$$\Rightarrow m + 1 = -5$$

$$\Rightarrow m = -6.$$

Therefore, the function f is differentiable at $x = 0$ for $m = -6$ and $n = 3$.

Example 5. Show that the function f defined by

$$f(x) = \begin{cases} x \dfrac{e^{\frac{1}{x}} - e^{-\frac{1}{x}}}{e^{\frac{1}{x}} + e^{-\frac{1}{x}}}, & x \neq 0 \\ 0, & x = 0, \end{cases}$$

is continuous, but not differentiable, at $x = 0$.

Solution. Here, we have

$$\text{L. H. L. } f(0-0) = \lim_{h \to 0} f(0-h) = \lim_{h \to 0} (-h) \left(\frac{e^{-\frac{1}{h}} - e^{\frac{1}{h}}}{e^{-\frac{1}{h}} + e^{\frac{1}{h}}} \right)$$

$$= \lim_{h \to 0} (-h) \left(\frac{e^{\frac{1}{h}} \left(e^{-\frac{2}{h}} - 1 \right)}{e^{\frac{1}{h}} \left(e^{-\frac{2}{h}} + 1 \right)} \right)$$

$$= \lim_{h \to 0} (-h) \left(\frac{e^{-\frac{2}{h}} - 1}{e^{-\frac{2}{h}} + 1} \right) = 0 \times \frac{0-1}{0+1} = 0,$$

because $\lim_{h \to 0} (e^{-\frac{2}{h}}) = 0$ and $\lim_{h \to 0} (h) = 0$,

$$\text{R. H. L. } f(0+0) = \lim_{h \to 0} f(0+h) = \lim_{h \to 0} (h) \left(\frac{e^{\frac{1}{h}} - e^{-\frac{1}{h}}}{e^{\frac{1}{h}} + e^{-\frac{1}{h}}} \right)$$

$$= \lim_{h \to 0} (h) \left(\frac{e^{\frac{1}{h}} \left(1 - e^{-\frac{2}{h}} \right)}{e^{\frac{1}{h}} \left(1 + e^{-\frac{2}{h}} \right)} \right)$$

$$= \lim_{h \to 0} (h) \left(\frac{1 - e^{-\frac{2}{h}}}{1 + e^{-\frac{2}{h}}} \right) = 0 \times \frac{1-0}{1+0} = 0,$$

and $f(0) = 0$.

Since $f(0-0) = f(0+0) = f(0)$, the function f is continuous at $x = 0$.

For differentiability at $x = 0$,

L. H. D. $f'(0 - 0) = \lim\limits_{h \to 0} \dfrac{f(0) - f(0 - h)}{h}$

$$= \lim\limits_{h \to 0} \left(\frac{0 - (-h)\left(\frac{e^{-\frac{1}{h}} - e^{\frac{1}{h}}}{e^{-\frac{1}{h}} + e^{\frac{1}{h}}} \right)}{h} \right) = \lim\limits_{h \to 0} \left(\frac{e^{-\frac{1}{h}} - e^{\frac{1}{h}}}{e^{-\frac{1}{h}} + e^{\frac{1}{h}}} \right)$$

$$= \lim\limits_{h \to 0} \left(\frac{e^{-\frac{2}{h}} - 1}{e^{-\frac{2}{h}} + 1} \right) = \frac{0 - 1}{0 + 1} = -1.$$

And

R. H. D. $f'(0 + 0) = \lim\limits_{h \to 0} \dfrac{f(0 + h) - f(0)}{h}$

$$= \lim\limits_{h \to 0} \left(\frac{(h)\left(\frac{e^{\frac{1}{h}} - e^{-\frac{1}{h}}}{e^{\frac{1}{h}} + e^{-\frac{1}{h}}} \right) - 0}{h} \right) = \lim\limits_{h \to 0} \left(\frac{e^{\frac{1}{h}} - e^{-\frac{1}{h}}}{e^{\frac{1}{h}} + e^{-\frac{1}{h}}} \right)$$

$$= \lim\limits_{h \to 0} \left(\frac{1 - e^{-\frac{2}{h}}}{1 + e^{-\frac{2}{h}}} \right) = \frac{1 - 0}{1 + 0} = 1.$$

Since $f'(0 - 0) \neq f'(0 + 0)$, the function f is not differentiable at $x = 0$.

So, the given function is continuous, but not differentiable, at $x = 0$.

Example 6. Show the function f defined by

$$f(x) = \begin{cases} x\dfrac{e^{\frac{1}{x}} - 1}{e^{\frac{1}{x}} + 1}, & x \neq 0, \\ 0, & x = 0, \end{cases}$$

is continuous, but not differentiable, at $x = 0$.

Solution. Here,

$$\text{L. H. L. } f(0-0) = \lim_{h \to 0} f(0-h) = \lim_{h \to 0} (-h) \left(\frac{e^{-\frac{1}{h}} - 1}{e^{-\frac{1}{h}} + 1} \right)$$

$$= 0 \times \frac{0-1}{0+1} = 0,$$

$$\text{R. H. L. } f(0+0) = \lim_{h \to 0} f(0+h) = \lim_{h \to 0} (h) \left(\frac{e^{\frac{1}{h}} - 1}{e^{\frac{1}{h}} + 1} \right)$$

$$= \lim_{h \to 0} (h) \left(\frac{e^{\frac{1}{h}} \left(1 - e^{-\frac{1}{h}} \right)}{e^{\frac{1}{h}} \left(1 + e^{-\frac{1}{h}} \right)} \right)$$

$$= \lim_{h \to 0} (h) \left(\frac{1 - e^{-\frac{1}{h}}}{1 + e^{-\frac{1}{h}}} \right) = 0 \times \frac{1-0}{1+0} = 0,$$

and $f(0) = 0$.

Since $f(0-0) = f(0+0) = f(0)$, the function f is continuous at $x = 0$.

For differentiability at $x = 0$, we have

$$\text{L. H. D. } f'(0-0) = \lim_{h \to 0} \frac{f(0) - f(0-h)}{h}$$

$$= \lim_{h \to 0} \left(\frac{0 - (-h) \left(\frac{e^{-\frac{1}{h}} - 1}{e^{-\frac{1}{h}} + 1} \right)}{h} \right)$$

$$= \lim_{h \to 0} \left(\frac{e^{-\frac{1}{h}} - 1}{e^{-\frac{1}{h}} + 1} \right) = \frac{0-1}{0+1} = -1$$

and

$$\text{R. H. D. } f'(0+0) = \lim_{h \to 0} \frac{f(0+h) - f(0)}{h} = \lim_{h \to 0} \left(\frac{(h) \left(\frac{e^{\frac{1}{h}} - 1}{e^{\frac{1}{h}} + 1} \right) - 0}{h} \right)$$

$$= \lim_{h \to 0} \left(\frac{e^{\frac{1}{h}} - 1}{e^{\frac{1}{h}} + 1} \right)$$

$$= \lim_{h \to 0} \left(\frac{1 - e^{-\frac{1}{h}}}{1 + e^{-\frac{1}{h}}} \right) = \frac{1 - 0}{1 + 0} = 1.$$

Since $f'(0-0) \neq f'(0+0)$, the function f is not differentiable, at $x = 0$.

So, the given function is continuous, but not differentiable, at $x = 0$.

7.2   Algebra of Derivatives

Theorem 1. *Let f and g be two functions defined on the closed interval $[a, b]$. If f and g are differentiable at $c \in [a, b]$, then their sum $f + g$ is also differentiable at c and $(f + g)'(c) = f'(c) + g'(c)$.*

Proof.   Since f is differentiable at $c \in [a, b]$, by the definition of differentiability, we get

$$\lim_{x \to c} \frac{f(x) - f(c)}{x - c} = f'(c). \tag{7.1}$$

Since g is differentiable at $c \in [a, b]$, by the definition of differentiability, we have

$$\lim_{x \to c} \frac{g(x) - g(c)}{x - c} = g'(c). \tag{7.2}$$

Now, we have

$$\lim_{x \to c} \frac{(f + g)(x) - (f + g)(c)}{x - c} = \lim_{x \to c} \frac{\{f(x) + g(x)\} - \{f(c) + g(c)\}}{x - c},$$

$$= \lim_{x \to c} \frac{\{f(x) - f(c)\} + \{g(x) - g(c)\}}{x - c}$$

$$= \lim_{x \to c} \frac{\{f(x) - f(c)\}}{x - c} + \lim_{x \to c} \frac{\{g(x) - g(c)\}}{x - c},$$

by Eqs. (7.1) and (7.2).

$$\Rightarrow \lim_{x \to c} \frac{(f + g)(x) - (f + g)(c)}{x - c} = f'(c) + g'(c).$$

Therefore, $f + g$ is differentiable at c and

$$(f + g)'(c) = f'(c) + g'(c).$$

Theorem 2. *Let f and g be two functions defined on the closed interval $[a, b]$. If f and g are differentiable at $c \in [a, b]$, then their difference $f - g$ is also differentiable at c and $(f - g)'(c) = f'(c) - g'(c)$.*

Proof. Since f is differentiable at $c \in [a, b]$, by the definition of differentiability, we have

$$\lim_{x \to c} \frac{f(x) - f(c)}{x - c} = f'(c). \tag{7.3}$$

Since g is differentiable at $c \in [a, b]$, by the definition of differentiability, we have

$$\lim_{x \to c} \frac{g(x) - g(c)}{x - c} = g'(c). \tag{7.4}$$

Now, we can write

$$\lim_{x \to c} \frac{(f - g)(x) - (f - g)(c)}{x - c} = \lim_{x \to c} \frac{\{f(x) - g(x)\} - \{f(c) - g(c)\}}{x - c},$$

$$= \lim_{x \to c} \frac{\{f(x) - f(c)\} - \{g(x) - g(c)\}}{x - c}$$

$$= \lim_{x \to c} \frac{\{f(x) - f(c)\}}{x - c} - \lim_{x \to c} \frac{\{g(x) - g(c)\}}{x - c},$$

by Eqs. (7.3) and (7.4).

$$\lim_{x \to c} \frac{(f - g)(x) - (f - g)(c)}{x - c} = f'(c) - g'(c).$$

Therefore, $f - g$ is differentiable at c and

$$(f - g)'(c) = f'(c) - g'(c).$$

Theorem 3. *Let f and g be two functions defined on the closed interval $[a, b]$. If f and g are derivable at $c \in [a, b]$, then their product fg is also differentiable at c and $(f.g)'(c) = f'(c)g(c) + f(c)g'(c)$.*

Proof.   Since f is differentiable at $c \in [a, b]$, it is also continuous at $c \in [a, b]$. Then, by the definition of continuity and differentiability of f at c, we have

$$\lim_{x \to c} f(x) = f(c) \quad \text{and} \quad \lim_{x \to c} \frac{f(x) - f(c)}{x - c} = f'(c). \qquad (7.5)$$

Since g is derivable at $c \in [a, b]$, it is also continuous at $c \in [a, b]$. Then, by the definition of continuity and differentiability of g at c, we get

$$\lim_{x \to c} g(x) = g(c) \quad \text{and} \quad \lim_{x \to c} \frac{g(x) - g(c)}{x - c} = g'(c). \qquad (7.6)$$

Now, to check the differentiability of the product function at $c \in [a, b]$ we consider

$$\lim_{x \to c} \frac{(fg)(x) - (fg)(c)}{x - c} = \lim_{x \to c} \frac{f(x)g(x) - f(c)g(c)}{x - c}$$

$$= \lim_{x \to c} \frac{f(x)g(x) - f(c)g(x) + f(c)g(x) - f(c)g(c)}{x - c},$$

$$= \lim_{x \to c} \frac{\{f(x) - f(c)\} g(x) + f(c) \{g(x) - g(c)\}}{x - c},$$

$$= \lim_{x \to c} \frac{\{f(x) - f(c)\}}{x - c} . \lim_{x \to c} g(x)$$

$$+ f(c) . \lim_{x \to c} \frac{\{g(x) - g(c)\}}{x - c},$$

by Eqs. (7.5) and (7.6).

$$\lim_{x \to c} \frac{(f + g)(x) - (f + g)(c)}{x - c} = f'(c).g(c) + f(c).g'(c).$$

Therefore, $f.g$ is differentiable at c and

$$(f.g)'(c) = f'(c).g(c) + f(c).g'(c).$$

Theorem 4. *Let f and g be two functions defined on the closed interval $[a, b]$. If f and g are differentiable at $c \in [a, b]$ and $g(c) \neq 0$, then $\frac{f}{g}$ is also differentiable at c and $\left(\frac{f}{g}\right)'(c) = \frac{\{f'(c).g(c) - f(c).g'(c)\}}{\{g(c)\}^2}$.*

Proof. Since f is differentiable at $c \in [a, b]$, it is also continuous at $c \in [a, b]$. Then, by the definition of continuity and differentiability of f at c, we have

$$\lim_{x \to c} f(x) = f(c) \quad \text{and} \quad \lim_{x \to c} \frac{f(x) - f(c)}{x - c} = f'(c). \qquad (7.7)$$

Since g is differentiable at $c \in [a, b]$, it is also continuous at $c \in [a, b]$. Then, by the definition of continuity and differentiability of g at c, we see that

$$\lim_{x \to c} g(x) = g(c) \quad \text{and} \quad \lim_{x \to c} \frac{g(x) - g(c)}{x - c} = g'(c). \qquad (7.8)$$

Now, to check the differentiability of product function at $c \in [a, b]$ consider

$$\lim_{x \to c} \frac{\left(\frac{f}{g}\right)(x) - \left(\frac{f}{g}\right)(c)}{x - c}$$

$$= \lim_{x \to c} \frac{f(x)g(c) - f(c)g(x)}{g(x)g(c).(x - c)}$$

$$= \lim_{x \to c} \frac{f(x)g(c) - f(c)\,g(c) + f(c)g(c) - f(c)g(x)}{g(x)g(c).(x - c)}$$

$$= \lim_{x \to c} \frac{\{f(x) - f(c)\}\,g(c) - f(c)\,\{g(x) - g(c)\}}{g(x)g(c).(x - c)}$$

$$= \frac{1}{g(c)} . \lim_{x \to c} \frac{1}{g(x)} \left\{ \lim_{x \to c} \frac{\{f(x) - f(c)\}}{x - c} . g(c) - f(c) \right.$$

$$\left. . \lim_{x \to c} \frac{\{g(x) - g(c)\}}{x - c} \right\},$$

by Eqs. (7.7) and (7.8).

$$\lim_{x \to c} \frac{(f + g)(x) - (f + g)(c)}{x - c} = \frac{1}{\{g(c)\}^2} \left\{ f'(c).g(c) - f(c).g'(c) \right\}.$$

Therefore, $\frac{f}{g}$ is differentiable at c and

$$\left(\frac{f}{g}\right)'(c) = \frac{\{f'(c).g(c) - f(c).g'(c)\}}{\{g(c)\}^2}.$$

7.3 Meaning of the Sign of the Derivative at an Interior Point

Let f be differentiable at $c \in [a, b]$ and suppose that its derivative $f'(c)$ exists and is positive at c. Then, by the definition of differentiability, we have

$$\lim_{x \to c} \frac{f(x) - f(c)}{x - c} = f'(c) > 0. \tag{7.9}$$

Thus, for each $\varepsilon > 0$, there exists a $\delta > 0$ such that

$$\left| \frac{f(x) - f(c)}{x - c} - f'(c) \right| < \varepsilon \quad \text{whenever } 0 < |x - c| < \delta.$$

That is,

$$f'(c) - \varepsilon < \frac{f(x) - f(c)}{x - c} < f'(c) + \varepsilon \quad \text{whenever } c - \delta < x < c + \delta \tag{7.10}$$

and $x \neq c$.

If we choose ε such that $0 < \varepsilon < f'(c)$ (we can find this because $f'(c) > 0$), then from Eq. (7.10) it is clear that

$$\frac{f(x) - f(c)}{x - c} > 0 \text{ whenever } c - \delta < x < c + \delta \quad \text{and} \quad x \neq c. \tag{7.11}$$

Now, if $x > c$, then $x - c > 0$. From Eq. (7.11), we get

$$f(x) - f(c) > 0.$$

Also, if $x < c$ then $x - c < 0$, then from Eq. (7.11), we have

$$f(x) - f(c) < 0.$$

So, from Eq. (7.11), we have the following two conclusions:

1. $f(x) - f(c) > 0$ or $f(x) > f(c)$, when $x > c$, that is, $x \in (c, c + \delta)$.
2. $f(x) - f(c) < 0$ or $f(x) < f(c)$ when $x < c$ that is, $x \in (c - \delta, c)$.

We conclude that $f(x)$ is a monotonic increasing function at $x = c$. So, if $f'(c) > 0$, then f is a monotonic increasing function at $x = c$.

Similarly, we can show that if $f'(c) < 0$, then f is a monotonic decreasing function at $x = c$.

7.4 Meaning of the Sign of the Derivative at Boundary Points

At boundary point $x = a$.

1. If $f'(a) > 0$, then $f(x) > f(a)$, when $x \in (a, a + \delta)$.
2. If $f'(a) < 0$, then $f(x) < f(a)$, when $x \in (a, a + \delta)$.

At boundary point $x = b$.

1. If $f'(b) > 0$, then $f(x) < f(b)$, when $x \in (b - \delta, b)$.
2. If $f'(b) < 0$, then $f(x) > f(b)$, when $x \in (b - \delta, b)$.

Example 1. Show that $\frac{x}{1+x} < \log(1 + x) < x, \ \forall \, x > 0$.

Solution. Let us define a function

$$g(x) = \log(1 + x) - \frac{x}{1 + x}. \tag{7.12}$$

Now, $g'(x) = \frac{1}{1+x} - \frac{1}{1+x} + \frac{x}{(1+x)^2}$.

$$\Rightarrow g'(x) = \frac{x}{(1 + x)^2} > 0, \quad \forall \, x > 0.$$

Therefore, $g(x)$ is a monotonic increasing function for all $x > 0$ and $g(0) = 0$.

So, if $x > 0$, then $g(x) > g(0) = 0$.

$$\Rightarrow g(x) > 0, \quad \forall \, x > 0.$$

By Eq. (7.12), we have

$$\log(1 + x) - \frac{x}{1 + x} > 0,$$

$$\Rightarrow \log(1 + x) > \frac{x}{1 + x}. \tag{7.13}$$

Let us define a new function

$$h(x) = x - \log(1 + x). \tag{7.14}$$

Now, $h'(x) = 1 - \frac{1}{1+x}$.

$$\Rightarrow h'(x) = \frac{x}{1 + x} > 0, \quad \forall \, x > 0.$$

Therefore, $h(x)$ is a monotonic increasing function for all $x > 0$ and $h(0) = 0$.

So, if $x > 0$, then $h(x) > h(0) = 0$.

$$\Rightarrow h(x) > 0, \quad \forall\, x > 0.$$

By Eq. (7.14),

$$x - \log(1 + x) > 0,$$
$$\Rightarrow \log(1 + x) < x. \tag{7.15}$$

From Eqs. (7.13) and (7.15), we get

$$\frac{x}{1 + x} < \log(1 + x) < x, \quad \forall\, x > 0.$$

7.5 Some Important Results

Darboux's theorem. *For a function f, which is differentiable on $[a, b]$, if $f'(a)$ and $f'(b)$ are opposite in sign, then there exists some $c \in (a, b)$ such that $f'(c) = 0$.*

Proof. Since $f'(a)$ and $f'(b)$ are opposite in sign, for simplicity we consider $f'(a) < 0$ and $f'(b) > 0$. Since $f'(a) < 0$, by the definition of the sign of the derivatives at the end points, there exists a positive number ε_1 such that

$$f(x) < f(a), \quad \forall\, x \in (a,\, a + \varepsilon_1). \tag{7.16}$$

Since $f'(b) > 0$, by the definition of the sign of the derivatives at the end points, there exists a positive number ε_2 such that

$$f(x) < f(b), \quad \forall\, x \in (b - \varepsilon_2,\, b). \tag{7.17}$$

Since f is differentiable on $[a, b]$, it is also continuous on $[a, b]$. Therefore, f attains its supremum and infimum on $[a, b]$. Let k be the infimum of f on $[a, b]$. Then there exists some $c \in [ab]$ such that

$$f(c) = k.$$

If $a = c$, then, by Eq. (7.16),

$$f(x) < f(c) = k, \quad \forall\, x \in (c,\, c + \varepsilon_1). \tag{7.18}$$

This contradicts the fact that k is the infimum of f on $[a, b]$. Therefore, $a \neq c$.

If $b = c$, then, by Eq. (7.17),

$$f(x) < f(c) = k, \quad \forall \, x \in (c - \varepsilon_2, c). \tag{7.19}$$

This contradicts the fact that k is the infimum of f on $[a, b]$. Therefore, $b \neq c$.

So, we conclude that $c \in (a, b)$. Now, it remains to prove that $f'(c) = 0$.

If $f'(c) > 0$, since c is an interior point, then, by the definition of the meaning of the sign of the derivatives, there exists a positive number ε_3 such that

$$f(x) < f(c) = k, \quad \forall \, x \in (c - \varepsilon_3, c). \tag{7.20}$$

which contradict the fact that k is the infimum of f on $[a, b]$.

If $f'(c) < 0$, since c is an interior point, then by the meaning of sign there exists a positive number ε_4 such that

$$f(x) < f(c) = k, \quad \forall \, x \in (c, c + \varepsilon_4). \tag{7.21}$$

This contradicts the fact that k is the infimum of f on $[a, b]$.

So, from Eqs. (7.20) and (7.21), we conclude that $f'(c) = 0$.

Intermediate value theorem for derivatives. *If the function f is differentiable on $[a, b]$, $f'(a) \neq f'(b)$ and m is any number lying between $f'(a)$ and $f'(b)$ then there exists some $c \in (a, b)$ such that $f'(c) = m$.*

Proof. Since $f'(a) \neq f'(b)$ and m is any number lying between $f'(a)$ and $f'(b)$ either

$$f'(a) < m < f'(b) \quad \text{or} \quad f'(b) < m < f'(a). \tag{7.22}$$

Let us define a new function

$$h(x) = f(x) - mx, \quad \forall \, x \in [ab]. \tag{7.23}$$

Since f is differentiable on $[a, b]$, by algebra of derivatives, h is also differentiable on $[a, b]$. Here, we have

$$h'(a) = f'(a) - m \quad \text{and} \quad h'(b) = f'(b) - m. \tag{7.24}$$

From Eqs. (7.22) and (7.24), it is clear that $h'(a)$ and $h'(b)$ are opposite in sign. By Darboux's theorem, for the function h there exists some $c \in (a, b)$ such that $h'(c) = 0$. So, from Eq. (7.23), we can write

$$h'(c) = f'(c) - m = 0,$$
$$\Rightarrow f'(c) = m.$$

Rolle's theorem. *If a function f defined on $[a, b]$ is*

i. *Continuous on $[a, b]$.*
ii. *Differentiable on (a, b)*
iii. *$f(a) = f(b)$,*

then there exists at least one real number $\alpha \in (a, b)$ such that $f'(\alpha) = 0$.

Proof. Since the function f is continuous on the closed interval $[a, b]$, it is bounded on $[a, b]$ and attains its supremum and infimum on $[a, b]$. Let K and k be the supremum and infimum of f respectively, on $[a, b]$ and let α and $\beta \in [a, b]$ be such that

$$f(\alpha) = K \quad \text{and} \quad f(\beta) = k.$$

Then there are two possibilities, either $K = k$ or $K \neq k$.

Case 1. If $K = k$, then $f(x) = K$, $\forall\, x \in [a, b]$.

That is, f is a constant function on $[a, b]$ and its derivative will be identically zero.

Therefore, $f'(x) = 0$, $\forall\, x \in [a, b]$. Thus, Rolle's theorem is true in this case.

Case 2. Suppose that $K \neq k$. As $f(a) = f(b)$ and $K \neq k$, at least one of the numbers K and k will be different from $f(a)$ and $f(b)$. Let K be this number. Then $f(\alpha) = K$, $K \neq f(a)$ and $K \neq f(b)$.

Now, $f(\alpha) \neq f(a) \Rightarrow \alpha \neq a$ and $f(\alpha) \neq f(b) \Rightarrow \alpha \neq b$.

This implies that α lies in the open interval (a, b).

It remains to prove that $f'(\alpha) = 0$.

If $f'(\alpha) > 0$, then there exists an interval $(c, c + \delta_1)$, with $\delta_1 > 0$, such that $f(x) > f(\alpha) = K$, $\forall\, x \in (c, c + \delta_1)$, which contradicts the fact that K is the supremum of f on $[a, b]$. Thus, we cannot have $f'(\alpha) > 0$.

If $f'(\alpha) < 0$, then there exists an interval $(c - \delta_2, c)$, with $\delta_2 > 0$, such that $f(x) > f(\alpha) = K$, $\forall\, x \in (c - \delta_2, c)$, which contradict the fact that K is the supremum of f on $[a, b]$. Thus, we cannot have $f'(\alpha) < 0$.

Hence, we have $f'(\alpha) = 0$ in this case as well.

Example 1. Using Rolle's theorem, find the value of α for the function $f(x) = \sin x$ in $[0, \pi]$.

Solution. Suppose that $f(x) = \sin x$ in $[0, \pi]$.

Then f is continuous on the closed interval $[0, \pi]$ and differentiable on the open interval $(0, \pi)$. Also

$$f(0) = f(\pi) = 0.$$

Hence, the function f fulfills all of the conditions of Rolle's theorem. So, by Rolle's theorem, there exists at least one $x \in (0, \pi)$, such that

$$f'(x) = 0 \Rightarrow \cos x = 0 \Rightarrow x = \frac{\pi}{2} \in (0, \pi).$$

Hence, in Rolle's theorem, $\alpha = \frac{\pi}{2}$.

Example 2. Discuss the applicability of Rolle's theorem to $f(x) = |x - 1|$ in $[0, 2]$.

Solution. Using the property of modulus, we can redefine the above function as follows:

$$f(x) = \begin{cases} 1 - x, & 0 \leq x < 1 \\ x - 1, & 1 \leq x \leq 2. \end{cases}$$

First, we will check its continuity at $\boldsymbol{x = 1}$.

L. H. L. $f(1 - 0) = \lim_{h \to 0} f(1 - h) = \lim_{h \to 0} (1 - (1 - h)) = \lim_{h \to 0} (h) = 0,$

R. H. L. $f(1 + 0) = \lim_{h \to 0} f(1 + h) = \lim_{h \to 0} (h) = 0,$

and $f(1) = 0$.

Since $f(1 - 0) = f(1 + 0) = f(1) = 0$, the function f is continuous at $x = 1$.

Consequently, f is continuous on $[0, 2]$ and the first condition of Rolle's theorem is verified. Now, for the differentiability of f at $\boldsymbol{x} = \boldsymbol{1}$

$$\text{L. H. D. } f'(1-0) = \lim_{h \to 0} \frac{f(1) - f(1-h)}{h} = \lim_{h \to 0} \left(\frac{0 - (1 - (1-h))}{h} \right)$$

$$= \lim_{h \to 0} \left(\frac{-h}{h} \right) = \lim_{h \to 0} (-1) = -1,$$

and

$$\text{R. H. D. } f'(1+0) = \lim_{h \to 0} \frac{f(1+h) - f(1)}{h} = \lim_{h \to 0} \left(\frac{(1+h) - 1 - 0}{h} \right)$$

$$= \lim_{h \to 0} \left(\frac{h}{h} \right) = \lim_{h \to 0} (1) = 1.$$

Since $f'(1-0) \neq f'(1+0)$, the function f is not differentiable at $x = 1 \in (0, 2)$. Thus, $f(x)$ is not differentiable in $(0, 2)$ and the second condition of Rolle's theorem is not satisfied. Hence, Rolle's theorem is not applicable to $f(x) = |x - 1|$ in $[0, 2]$.

Example 3. Show that, for any real number m, the polynomial given by $x^3 + x + m$ has exactly one real root.

Solution. Since $p(x) = x^3 + x + m$ is an odd-degree polynomial, so it must have one real root. If possible, let it have two distinct real roots a and b. The polynomial $p(x) = x^3 + x + m$ is continuous on the closed interval $[a, b]$. It is also differentiable on (a, b). Since a and b are the roots of polynomial $p(x) = x^3 + x + m$, we have

$$p(a) = p(b) = 0.$$

All of the conditions of Rolle's theorem are verified. Therefore, there exists at least one real number $\alpha \in (a, b)$ such that

$$p'(\alpha) = 0.$$
$$\Rightarrow 3\alpha^2 + 1 = 0.$$

The above equation contradicts the fact that α is a real number, there by contradicting Rolle's theorem. So our assumption for the existence of two distinct real numbers is not possible. Therefore, for any real number m, the polynomial given by $x^3 + x + m$ has exactly one real root.

Lagrange's mean value theorem. *If a function f defined on $[a, b]$ is*

i. *Continuous on $[a, b]$.*
ii. *Differentiable on (a, b),*

then there exists at least one real number $\alpha \in (a, b)$ such that $\frac{f(b)-f(a)}{b-a} = f'(\alpha)$.

Proof. Define a function h as follows:

$$h(x) = f(x) + \lambda x,$$

where λ is such that $h(a) = h(b)$.
Therefore, $f(a) + \lambda a = f(b) + \lambda b$.

$$\Rightarrow \lambda = -\frac{f(b) - f(a)}{b - a}. \tag{7.25}$$

Since the function $h(x)$ is the sum of two continuous and differentiable functions, so $h(x)$ is itself

a. Continuous on $[a, b]$
b. Differentiable on (a, b)
c. $h(a) = h(b)$.

Then, by Rolle's theorem, there exists at least one real number $\alpha \in (ab)$ such that $h'(\alpha) = 0$.

$$\Rightarrow h'(\alpha) = f'(\alpha) + \lambda = 0.$$
$$\Rightarrow \lambda = -f'(\alpha) \tag{7.26}$$

From Eqs. (7.25) and (7.26), we get

$$\frac{f(b) - f(a)}{b - a} = f'(\alpha).$$

Cauchy's mean value theorem. *Suppose that two functions f and g defined on $[a, b]$ are*

i. *Continuous on* $[a, b]$
ii. *Differentiable on* (a, b)
iii. $g'(x) \neq 0$, *for any* $x \in (a, b)$.

Then there exists at least one real number $\alpha \in (a, b)$ *such that* $\frac{f(b)-f(a)}{g(b)-g(a)} = \frac{f'(\alpha)}{g'(\alpha)}$.

Proof. Define a function h as follows:

$$h(x) = f(x) + \lambda g(x),$$

where λ is such that $h(a) = h(b)$.

Therefore, $f(a) + \lambda g(a) = f(b) + \lambda g(b)$.

If $g(a) = g(b)$, then the function $g(x)$ satisfies all of the conditions of Rolle's theorem and hence its derivative will vanish at least once in (a, b) and the condition (iii) would be violated. Therefore, $g(b) - g(a) \neq 0$. On this account, from the above we can write

$$\lambda = -\frac{f(b) - f(a)}{g(b) - g(a)}. \tag{7.27}$$

Since the function $h(x)$ is the sum of two continuous and differentiable functions, $h(x)$ is itself

a. Continuous on $[a, b]$
b. Differentiable on (a, b)
c. $h(a) = h(b)$.

Then, by Rolle's theorem, there exists at least one real number $\alpha \in (ab)$ such that $h'(\alpha) = 0$.

$$\Rightarrow h'(\alpha) = f'(\alpha) + \lambda g'(\alpha) = 0.$$

$$\Rightarrow \lambda = -\frac{f'(\alpha)}{g'(\alpha)}, \text{ because } (g'(\alpha) \neq 0). \tag{7.28}$$

From Eqs. (7.27) and (7.28), we get

$$\frac{f(b) - f(a)}{g(b) - g(a)} = \frac{f'(\alpha)}{g'(\alpha)}.$$

Exercises

1. Show that the function $f(x) = |x| + |x - 1|, \forall x \in \mathbf{R}$ is differentiable everywhere except at the points $x = 0$ and $x = 1$.
2. Show that the function $f(x) = |x - 2| + |x + 2|, \forall x \in \mathbf{R}$, is differentiable everywhere except at the points $x = -2$ and $x = 2$.
3. Show that the function $f(x) = |x - 2| + |x| + |x + 2|, \forall x \in \mathbf{R}$, is differentiable everywhere except at the points $x = -2$, $x = 0$, and $x = 2$.
4. Discuss the continuity and differentiability of the following function at $x = 1$.

$$f(x) = \begin{cases} 1, & 0 < x \leq 1 \\ \frac{1}{x}, & x > 1. \end{cases}$$

   Ans. Continuous, but not differentiable, at $x = 1$.
5. Show that the following function f is continuous and differentiable at $x = 0$.

$$f(x) = \begin{cases} x^2 \sin\left(\frac{1}{x}\right), & x \neq 0 \\ 0, & x = 0. \end{cases}$$

6. Determine the values of n for which the following function f is continuous and differentiable at $x = 0$

$$f(x) = \begin{cases} x^n \cos\left(\frac{1}{x}\right), & x \neq 0 \\ 0, & x = 0. \end{cases}$$

   Ans. Continuous for $n > 0$ and differentiable for $n > 1$.
7. Find the values of the constants m and n for which the following function f is differentiable at $x = 1$.

$$f(x) = \begin{cases} x^2 + 3x + m, & x \leq 1 \\ nx + 2, & x > 1. \end{cases}$$

   Ans. $m = 3$, $n = 5$.
8. Show the function f defined by

$$f(x) = \begin{cases} \dfrac{x}{e^{\frac{1}{x}} + 1}, & x \neq 0, \\ 0, & x = 0, \end{cases}$$

   is continuous, but not differentiable, at $x = 0$.
9. Show that $\frac{x}{1+x^2} < \tan^{-1} x < x, \forall x > 0$.

Chapter 8

Uniform Convergence of Sequences and Series of Real Functions

In Chapter 4, we considered the real sequences whose terms are real numbers. In this chapter, we consider the sequence whose terms are real functions rather than real numbers. These sequences naturally arise in Real Analysis. The sequences of functions are mainly useful in approximating some given function and, by using this approximation, we can define new functions from the known one.

In this chapter, first we define two notions of convergence of sequences of functions: point-wise convergence and uniform convergence. We discuss some results for the test of convergence of sequences of functions. Further, we introduce the infinite series of functions due to their frequent appearance and importance. In the next section, we define the notions of point-wise and uniform convergence of a series of functions. Finally, we list some important theorems such the Weierstrass M-test, the Abel test, and the Dirichlet test for the uniform convergence of a series of functions.

8.1 Point-Wise Convergence of Sequences of Functions

Let $\langle f_n \rangle$ be a sequence of functions defined on $[a, b]$. Suppose that $x_1 \in [a, b]$ is any arbitrary point. Then there corresponds a sequence $\langle f_n(x_1) \rangle$ of real numbers with the terms $f_1(x_1), f_2(x_1), f_3(x_1), \ldots, f_n(x_1), \ldots$.

Let $x_2 \in [a, b]$ be any arbitrary point. Then there corresponds a sequence of $\langle f_n(x_2) \rangle$ of real numbers with the terms $f_1(x_2), f_2(x_2), f_3(x_2), \ldots, f_n(x_2), \ldots$.

Therefore, for each $x \in [a, b]$, we can find a sequence $\langle f_n(x) \rangle$ of real numbers with the terms $f_1(x), f_2(x), f_3(x), \ldots, f_n(x), \ldots$.

Suppose that the sequence $\langle f_n(x) \rangle$ of real numbers converges for each $x \in [a, b]$. Let the sequence $\langle f_n(x_1) \rangle$ converge to $f(x_1)$, the sequence $\langle f_n(x_2) \rangle$ converges to $f(x_2), \ldots$, and so on.

Then the new function f defined on $[a, b]$ is said to be the point-wise limit of the sequence $\langle f_n \rangle$ of functions on $[a, b]$.

Thus, if the function f is the point-wise limit of the sequence $\langle f_n \rangle$ of functions on $[a, b]$, then, for each $\varepsilon > 0$ and for each $x \in [a, b]$, there exists a natural number $m(\varepsilon, x)$ (depending on both x and ε) such that

$$|f_n(x) - f(x)| < \varepsilon, \quad \forall \, n \geq m.$$

Remark. If we want to check that the sequence $\langle f_n \rangle$ of functions defined on $[a, b]$ is point-wise convergent at some $x \in [a, b]$, then we will calculate the limit of the function $f_n(x)$, that is, $\lim_{n \to \infty} f_n(x)$. If the limit is finite, then the sequence $\langle f_n \rangle$ of functions is point-wise convergent at $x \in [a, b]$. If the limit is not finite, then the sequence $\langle f_n \rangle$ of functions is not point-wise convergent at $x \in [a, b]$.

Example 1. Consider the sequence of functions $\langle f_n \rangle$ on $[-2, 2]$ defined by

$$f_n(x) = \frac{x}{n^2}, \quad \forall \, x \in [-2, 2].$$

For the point-wise convergence, we have

$$f(x) = \lim_{n \to \infty} f_n(x) = \lim_{n \to \infty} \frac{x}{n^2} = 0, \quad \forall \, x \in [-2, 2].$$

That is, $f(x) = 0, \forall x \in [-2, 2]$ is the point-wise limit of the sequence $\langle \frac{x}{n^2} \rangle$ of functions.

Example 2. Consider the sequence $\langle f_n \rangle$ of functions on $[0, 1]$ defined by

$$f_n(x) = nx, \quad \forall \, x \in [0, 1].$$

For the point-wise convergence, consider $x = 0$, so that

$$f(x) = \lim_{n \to \infty} f_n(x) = \lim_{n \to \infty} nx = 0.$$

Consider $0 < x \leq 1$, in which case we have

$$f(x) = \lim_{n \to \infty} f_n(x) = \lim_{n \to \infty} nx = \infty \quad \text{for } 0 < x \leq 1.$$

So, the sequence $\langle nx \rangle$ is point-wise convergent only at $x = 0$ and the point-wise limit is 0.

8.2   Uniform Convergence of Sequences of Functions

If the function f is the uniform limit of the sequence $\langle f_n \rangle$ of functions on $[a, b]$, then, for each $\varepsilon > 0$ and for all $x \in [a, b]$, there exists a natural number $m(\varepsilon)$ (depending on ε only) such that

$$|f_n(x) - f(x)| < \varepsilon, \quad \forall\, x \in [a, b] \quad \text{and} \quad \forall\, n \geq m.$$

Remark. In point-wise convergence, the value of m depends on both x and ε, whereas in the uniform convergence the value of m depends on ε only.

Theorem. *Let $\langle f_n \rangle$ be the sequence of functions which converges uniformly on $[a, b]$. Then it converges point-wise on $[a, b]$, but the converse is not true.*

Proof. From the definitions of the point-wise and uniform convergence, it is clear that every uniformly convergent sequence of functions is also point-wise convergent.

To show that the converse is not true, we will give an example of a sequence of functions which is point-wise convergent but not uniform convergent.

Define a sequence of functions as follows:

$$f_n(x) = \frac{nx}{1 + n^2 x^2}, \quad \forall x \in [-1, 1].$$

For point-wise convergence, we have

$$f(x) = \lim_{n \to \infty} f_n(x) = \lim_{n \to \infty} \frac{nx}{1 + n^2 x^2} = 0, \quad \forall x \in [-1, 1].$$

$$f(x) = 0, \quad \forall x \in [-1, 1].$$

Therefore, the sequence $\langle f_n \rangle$ of functions converges point-wise on $[-1, 1]$ and the point-wise limit is 0.

Let the sequence $\langle f_n \rangle$ of functions converge uniformly on $[-1, 1]$. Then the point-wise limit will also be the uniform limit. Therefore, for $\varepsilon > 0$, there exists a natural number m such that

$$\left| \frac{nx}{1 + n^2 x^2} - 0 \right| = \frac{nx}{1 + n^2 x^2} < \varepsilon, \ \forall x \in [-1, 1], \quad \text{and} \quad \forall \, n \geq m. \tag{8.1}$$

Let m_1 be a natural number greater than m such that $\frac{1}{m_1} \in [-1, 1]$. Consider $\varepsilon = \frac{1}{4}$ and, in particular, take $x = \frac{1}{m_1} \in [-1, 1]$. Then, by Eq. (8.1) for $n = m_1 \geq m$, we have

$$\frac{m_1 \times \frac{1}{m_1}}{1 + m_1^2 \left(\frac{1}{m_1} \right)^2} = \frac{1}{2} > \frac{1}{4} = \varepsilon,$$

which contradicts the fact that $\langle f_n \rangle$ converges uniformly on $[-1, 1]$. So, our assumption is false. Therefore, the sequence $\langle f_n \rangle$ does not converge uniformly on $[-1, 1]$.

Example 1. Examine the uniform convergence of the sequence $\langle f_n \rangle$ of functions defined by $f_n(x) = \frac{x}{n+x}$ on the interval $[0, l]$ and $[0, \infty)$.

Solution. For the point-wise convergence

$$f(x) = \lim_{n \to \infty} f_n(x) = \lim_{n \to \infty} \frac{x}{n + x} = 0, \quad \forall \, x > 0,$$

and

$$f(0) = \lim_{n \to \infty} f_n(0) = \lim_{n \to \infty} \frac{0}{n + 0} = 0.$$

$$\Rightarrow f(x) = 0, \quad \forall \, x \in [0, \infty).$$

Therefore, the sequence $\langle \frac{x}{n+x} \rangle$ of functions converges point-wise on $[0, l]$ and $[0, \infty)$ and the point-wise limit is 0.

For uniform convergence, let $\varepsilon > 0$ be given and $0 < x \leq l$. Then

$$\left| \frac{x}{n+x} - 0 \right| = \frac{x}{n+x} < \varepsilon,$$

$$\text{if } \frac{n+x}{x} > \frac{1}{\varepsilon},$$

$$\text{or} \quad \text{if } n > \left(\frac{1}{\varepsilon} - 1 \right) x.$$

The number $(\frac{1}{\varepsilon} - 1)x$ increases with the increase of x, and the maximum value of this number on interval $[0, l]$ is $(\frac{1}{\varepsilon} - 1)l$.

Let m be a natural number with $m \geq (\frac{1}{\varepsilon} - 1)l$. Then

$$|f_n(x) - f(x)| < \varepsilon, \quad \forall \, 0 < x \leq l, \quad \text{and} \quad \forall n \geq m.$$

At $x = 0$,

$$|f_n(x) - f(x)| = 0 < \varepsilon, \quad \forall \, n \geq 1.$$

So, the given function is uniformly convergent on $[0, l]$.

Since the number $(\frac{1}{\varepsilon} - 1)x \to \infty$ as $x \to \infty$, it is not possible to find a natural number m such that

$$|f_n(x) - f(x)| < \varepsilon, \quad \forall \, x \in [0, \infty), \quad \text{and} \quad \forall \, n \geq m.$$

So, the given function is not uniformly convergent on $[0, \infty)$.

Example 2. Show that the sequence $\langle f_n \rangle$ of functions defined by $f_n(x) = e^{-nx}$ is uniformly convergent in $[a, b]$, where $a, b > 0$, but only point-wise convergent on $[0, b]$.

Solution. For point-wise convergence, we have

$$f(x) = \lim_{n \to \infty} f_n(x) = \lim_{n \to \infty} e^{-nx} = 0, \quad \forall \, x \in [a, b],$$

and

$$f(0) = \lim_{n \to \infty} f_n(0) = \lim_{n \to \infty} 1 = 1.$$

Therefore, the sequence $\langle e^{-nx} \rangle$ of functions converges point-wise on $[0, b]$ and $[a, b]$, where $a, b > 0$.

For uniform convergence, let $\varepsilon > 0$ be given and $a \leq x \leq b$. Then

$$|e^{-nx} - 0| = e^{-nx} < \varepsilon, \quad \text{if } n > \frac{\log\left(\frac{1}{\varepsilon}\right)}{x \log e}.$$

The number $\frac{\log(\frac{1}{\varepsilon})}{x \log e}$ increases with the decrease of x, and the maximum value of this number on the interval $[a, b]$ is $\frac{\log(\frac{1}{\varepsilon})}{a \log e}$.

Let m be a natural number with $m \geq \frac{\log(\frac{1}{\varepsilon})}{a \log e}$. Then

$$|f_n(x) - f(x)| < \varepsilon, \quad \forall\, x \in [a, b], \quad \text{and} \quad \forall\, n \geq m.$$

So the given sequence of functions is uniformly convergent on $[a, b]$.

Since the number $\frac{\log(\frac{1}{\varepsilon})}{x \log e} \to \infty$ as $x \to 0$, it is not possible to find a natural number m such that

$$|f_n(x) - f(x)| < \varepsilon, \quad \forall\, x \in [0, b], \quad \text{and} \quad \forall\, n \geq m.$$

So the given function is not uniformly convergent on $[0, b]$.

Example 3. Show that the sequence $\langle f_n \rangle$ of functions defined by $f_n(x) = x^n$ is uniformly convergent in $[0, l]$, where $l < 1$, but only point-wise convergent on $[0, 1]$.

Solution. For point-wise convergence, we have

$$f(x) = \lim_{n\to\infty} f_n(x) = \lim_{n\to\infty} x^n = 0, \quad \forall 0 < x < 1,$$

$$f(0) = \lim_{n\to\infty} f_n(0) = \lim_{n\to\infty} 0 = 0,$$

and

$$f(1) = \lim_{n\to\infty} f_n(1) = \lim_{n\to\infty} 1 = 1.$$

$$\Rightarrow f(x) = \begin{cases} 0, & 0 \leq x < 1 \\ 1, & x = 1. \end{cases}$$

Therefore, the sequence $\langle x^n \rangle$ of functions converges point-wise on both the intervals $[0, l]$, where $l < 1$, and $[0, 1]$.

For uniform convergence, let $\varepsilon > 0$ be given and $0 < x \le l$. Then

$$|x^n - 0| = x^n < \varepsilon, \quad \text{if } n \log\left(\frac{1}{x}\right) > \log\left(\frac{1}{\varepsilon}\right),$$

or if $n > \dfrac{\log\left(\frac{1}{\varepsilon}\right)}{\log\left(\frac{1}{x}\right)}$.

Since the number $\dfrac{\log\left(\frac{1}{\varepsilon}\right)}{\log\left(\frac{1}{x}\right)}$ increases with x, it attains its maximum value on the interval $[0, l]$, as $\dfrac{\log\left(\frac{1}{\varepsilon}\right)}{\log\left(\frac{1}{l}\right)}$.

Let m be a natural number with $m \ge \dfrac{\log\left(\frac{1}{\varepsilon}\right)}{\log\left(\frac{1}{l}\right)}$. Then

$$|f_n(x) - f(x)| < \varepsilon, \quad \forall 0 < x \le l \quad \text{and} \quad \forall n \ge m.$$

At $x = 0$,

$$|f_n(x) - f(x)| = 0 < \varepsilon, \quad \forall n \ge 1.$$

So, the given function is uniformly convergent on $[0, l]$.

Since the number $\dfrac{\log\left(\frac{1}{\varepsilon}\right)}{\log\left(\frac{1}{x}\right)} \to \infty$ as $x \to 1$, it is not possible to find a natural number m such that

$$|f_n(x) - f(x)| < \varepsilon, \quad \forall\, x \in [0, 1], \quad \text{and} \quad \forall n \ge m.$$

So, the given sequence of functions is not uniformly convergent on $[0, 1]$.

8.3   Test for Uniform Convergence of Sequences of Functions

Theorem (Weierstrass's M-test). *Let $\langle f_n \rangle$ be a sequence of functions defined on $[a, b]$ such that*

$$\lim_{n \to \infty} f_n(x) = f(x), \quad \forall\, x \in [a, b],$$

and assume that $M_n = \sup_{x \in [a,b]} |f_n(x) - f(x)|$.

Then $f_n \to f$ uniformly if and only if $M_n \to 0$ as $n \to \infty$.

Proof. Let the sequence $\langle f_n \rangle$ converge uniformly to f on $[a, b]$. Then, for each $\varepsilon > 0$, there exists a natural number m independent of the point x such that

$$|f_n(x) - f(x)| < \varepsilon, \quad \forall\, n \geq m, \quad \forall\, x \in [a, b].$$

$$\Rightarrow \sup |f_n(x) - f(x)| < \varepsilon, \quad \forall\, n \geq m, \quad \forall\, x \in [a, b].$$

$$\Rightarrow M_n = \sup_{x \in [a,b]} |f_n(x) - f(x)| < \varepsilon, \quad \forall n \geq m,$$

$$(\text{by the definition of } M_n).$$

$$\Rightarrow M_n = |M_n - 0| < \varepsilon, \quad \forall\, n \geq m, \text{ (since } M_n \text{ is positive).}$$

$$\Rightarrow M_n \to 0 \text{ as } n \to \infty.$$

Conversely, let $M_n \to 0$ as $n \to \infty$. Then we will show that the sequence $\langle f_n \rangle$ converges uniformly to f on $[a, b]$. Since $M_n \to 0$ as $n \to \infty$, for each $\varepsilon > 0$, there exists a natural number m such that

$$M_n = |M_n - 0| < \varepsilon, \quad \forall\, n \geq m, \text{ (since } M_n \text{ is positive).}$$

$$\Rightarrow M_n = \sup_{x \in [a,b]} |f_n(x) - f(x)| < \varepsilon, \quad \forall\, n \geq m,$$

$$(\text{by the definition of } M_n).$$

$$\Rightarrow \sup |f_n(x) - f(x)| < \varepsilon, \quad \forall\, n \geq m, \forall\, x \in [a, b].$$

$$\Rightarrow |f_n(x) - f(x)| < \varepsilon, \quad \forall\, n \geq m, \quad \forall\, x \in [a, b].$$

$$\Rightarrow f_n \text{ converges uniformly to } f \text{ on } [a, b].$$

Example 1. Show that the sequence $\langle f_n \rangle$, where $f_n(x) = \frac{x}{n(1+nx^2)}$, is uniformly convergent for all $x \geq 0$.

Solution. The limit of the sequence of functions is given by

$$f(x) = \lim_{n \to \infty} f_n(x) = \lim_{n \to \infty} \frac{x}{n(1 + nx^2)} = 0, \quad \forall\, x > 0.$$

At $x = 0$,

$$f(x) = \lim_{n \to \infty} f_n(x) = \lim_{n \to \infty} 0 = 0.$$

Therefore, we have

$$f(x) = 0 \quad \forall\, x \geq 0.$$

Now, it is seen that

$$|f_n(x) - f(x)| = \left| \frac{x}{n(1 + nx^2)} - 0 \right| = \left| \frac{x}{n(1 + nx^2)} \right| = \frac{x}{n(1 + nx^2)}.$$

Let

$$y = \frac{x}{n(1 + nx^2)}.$$

We will calculate maximum value of y.

$$\frac{dy}{dx} = \frac{(1 - nx^2)}{n(1 + nx^2)^2},$$

and, for the maxima and minima,

$$\frac{dy}{dx} = \frac{(1 - nx^2)}{n(1 + nx^2)^2} = 0.$$

$$\Rightarrow x = \pm \frac{1}{\sqrt{n}}, \text{ since } x \geq 0,$$

so we will check the maximum value only at

$$x = \frac{1}{\sqrt{n}}.$$

We have

$$\frac{d^2y}{dx^2} = \frac{2x}{(1 + nx^2)^3}(-3 + nx^2),$$

at $x = \frac{1}{\sqrt{n}}$, $\frac{d^2y}{dx^2} = -\frac{1}{2\sqrt{n}} < 0$.

So, y attains its maximum value at $x = \frac{1}{\sqrt{n}}$, which is given as $\frac{1}{2n^{3/2}}$.

Now,

$$M_n = \sup_{x \geq 0} |f_n(x) - f(x)| = \text{maximum value of } y = \frac{1}{2n^{3/2}}.$$

That is,

$$M_n = \frac{1}{2n^{3/2}} \quad \text{and} \quad \lim_{n \to \infty} M_n = \lim_{n \to \infty} \frac{1}{2n^{3/2}} = 0.$$

So, by the Weierstrass M-test, the sequence $\left\langle \frac{x}{n(1+nx^2)} \right\rangle$ of functions is uniformly convergent for all $x \geq 0$.

Example 2. Show that the sequence $\langle f_n \rangle$, where $f_n(x) = \frac{n^2 x}{(1 + n^3 x^2)}$, is not uniformly convergent on $[0, 1]$.

Solution. The limit of this sequence of functions is given by

$$f(x) = \lim_{n \to \infty} f_n(x) = \lim_{n \to \infty} \frac{n^2 x}{(1 + n^3 x^2)} = 0, \quad \forall 0 < x \le 1.$$

At $x = 0$,

$$f(x) = \lim_{n \to \infty} f_n(x) = \lim_{n \to \infty} 0 = 0.$$

Therefore, we have

$$f(x) = 0 \quad \forall x \in [0, 1].$$

Now, we see that

$$|f_n(x) - f(x)| = \left| \frac{n^2 x}{(1 + n^3 x^2)} - 0 \right| = \left| \frac{n^2 x}{(1 + n^3 x^2)} \right| = \frac{n^2 x}{(1 + n^3 x^2)}.$$

Let

$$y = \frac{n^2 x}{(1 + n^3 x^2)}.$$

We will calculate the maximum value of y.

$$\frac{dy}{dx} = \frac{n^2 (1 - n^3 x^2)}{(1 + n^3 x^2)^2},$$

and, for the maxima and minima,

$$\frac{dy}{dx} = \frac{n^2 (1 - n^3 x^2)}{(1 + n^3 x^2)^2} = 0.$$

$$\Rightarrow x = \pm \frac{1}{n^{3/2}}, \text{ since } x \in [0, 1],$$

so we will check the maximum value only at $x = \frac{1}{n^{3/2}}$.
We have

$$\frac{d^2 y}{dx^2} = \frac{2n^5 x}{(1 + n^3 x^2)^3}(-3 + n^3 x^2),$$

at

$$x = \frac{1}{n^{3/2}}, \frac{d^2 y}{dx^2} = -\frac{1}{2} n^{7/2} < 0.$$

So, y attains its maximum value at $x = \frac{1}{n^{3/2}}$, which is given as $\frac{\sqrt{n}}{2}$.

Now,

$$M_n = \sup_{x \in [0,1]} |f_n(x) - f(x)| = \text{maximum value of } y = \frac{\sqrt{n}}{2}.$$

That is,

$$M_n = \frac{\sqrt{n}}{2} \quad \text{and} \quad \lim_{n \to \infty} M_n = \lim_{n \to \infty} \frac{\sqrt{n}}{2} = \infty.$$

So, by the Weierstrass M-test, the sequence of functions $\left\langle \frac{n^2 x}{(1+n^3 x^2)} \right\rangle$ is not uniformly convergent on $[0, 1]$.

Example 3. Show that the sequence $\langle f_n \rangle$, where $f_n(x) = nxe^{-nx^3}$, is not uniformly convergent for all $x \geq 0$.

Solution. The limit of the sequence of functions is given by

$$f(x) = \lim_{n \to \infty} f_n(x) = \lim_{n \to \infty} nxe^{-nx^3} = 0 \quad \forall x > 0.$$

At $x = 0$,

$$f(x) = \lim_{n \to \infty} f_n(x) = \lim_{n \to \infty} 0 = 0.$$

Therefore, we have

$$f(x) = 0 \quad \forall x \geq 0.$$

Now, $|f_n(x) - f(x)| = |nxe^{-nx^3} - 0| = |nxe^{-nx^3}| = nxe^{-nx^3}.$
Let

$$y = nxe^{-nx^3}.$$

Then we will calculate maximum value of y.

$$\frac{dy}{dx} = ne^{-nx^3}(1 - 3nx^3),$$

so, for the maxima and minima,

$$\frac{dy}{dx} = ne^{-nx^3}(1 - 3nx^3) = 0$$

$$\Rightarrow x = \frac{1}{(3n)^{1/3}}.$$

Further,

$$\frac{d^2y}{dx^2} = -3n^2x^2e^{-nx^3}(4 - 3nx^3),$$

at

$$x = \frac{1}{(3n)^{1/3}}, \frac{d^2y}{dx^2} = -\frac{\sqrt{3n}}{e^{1/3}} < 0.$$

So, y attains its maximum value at $x = \frac{1}{(3n)^{1/3}}$,which is given as $\sqrt[3]{\frac{n^2}{3e}}$.

Now,

$$M_n = \sup_{x \geq 0}|f_n(x) - f(x)| = \text{maximum value of } y = \sqrt[3]{\frac{n^2}{3e}}.$$

That is,

$$M_n = \sqrt[3]{\frac{n^2}{3e}} \quad \text{and} \quad \lim_{n \to \infty} M_n = \lim_{n \to \infty} \sqrt[3]{\frac{n^2}{3e}} = \infty.$$

So. by the Weierstrass M-test, the sequence $\langle nxe^{-nx^3} \rangle$ of functions is not uniformly convergent for all $x \geq 0$.

8.4 Point-Wise and Uniform Convergence of Series of Functions

If the series $\sum f_n$ of functions converges for every $x \in [a, b]$, we define

$$f(x) = \sum_{n=0}^{\infty} f_n(x).$$

Then the function f is called the point-wise sum of the series $\sum f_n$ on $[a, b]$.

A series $\sum f_n$ of functions converges uniformly on $[a, b]$ if the sequence $\langle s_n \rangle$ of its partial sums defined by

$$s_n(x) = \sum_{r=1}^{n} f_r(x)$$

converges uniformly on $[a, b]$.

Thus, a series $\sum f_n$ of functions converges uniformly to the function f on $[a, b]$, if for each $\varepsilon > 0$ and for all $x \in [a, b]$, there exists a natural number $m(\varepsilon)$ (depending on ε only) such that

$$|s_n(x) - f(x)| < \varepsilon, \ \forall x \in [a, b], \quad \text{and} \quad \forall n \geq m.$$

That is,

$$|f_1(x) + f_2(x) + \cdots + f_n(x) - f(x)| < \varepsilon, \quad \forall x \in [a, b], \quad \text{and} \quad \forall n \geq m.$$

Now, we discuss some important theorems to test the uniform convergence of a series of functions.

Theorem 1 (Weierstrass's M-test for series of functions).
A series $\sum u_n$ of functions converges uniformly in $[a, b]$ if there exists a series of positive terms $\sum M_n$ such that

$$|u_n(x)| \leq M_n, \quad \forall \ x \in [a, b].$$

Proof. Since the series $\sum M_n$ of positive terms is convergent, for each $\varepsilon > 0$, there exists a natural number m such that

$$|M_{n+1} + M_{n+2} + M_{n+3} + \cdots + M_{n+p}| < \varepsilon,$$
$$\forall \ n \geq m \quad \text{and} \quad p \geq 1. \tag{8.2}$$

Given the condition:

$$|u_n(x)| \leq M_n, \quad \forall \ x \in [a, b]. \tag{8.3}$$

We see that

$$|u_{n+1}(x) + u_{n+2}(x) + u_{n+3}(x) + \cdots + u_{n+p}(x)|$$
$$\leq |u_{n+1}(x)| + |u_{n+2}(x)| + |u_{n+3}(x)| + \cdots + |u_{n+p}(x)|.$$

Using Eq. (8.3) in the above expression, we get

$$|u_{n+1}(x) + u_{n+2}(x) + u_{n+3}(x) + \cdots + u_{n+p}(x)|$$
$$\leq M_{n+1} + M_{n+2} + M_{n+3} + \cdots + M_{n+p} < \varepsilon.$$

Now, using Eq. (8.2), we have

$$\Rightarrow |u_{n+1}(x) + u_{n+2}(x) + u_{n+3}(x) + \cdots + u_{n+p}(x)| < \varepsilon,$$
$$\forall \ x \in [a, b], \quad \forall \ n \geq m, \quad \text{and} \quad p \geq 1.$$

Therefore, the series $\sum u_n$ of functions converges uniformly in $[a, b]$.

Example 1. Using the Weierstrass M-test, examine the uniform convergence of $\sum \frac{\sin nx}{n^2}$ and $\sum \frac{\sin nx}{n}$ in $\mathbf{R}$.

Solution. Since

$$\left| \frac{\sin nx}{n^2} \right| \le \frac{1}{n^2}, \quad \forall \, x \in \mathbf{R},$$

we take $M_n = \frac{1}{n^2}$. By the p-test, the series $\sum \frac{1}{n^2}$ is convergent because $p = 2$. Therefore, by the Weierstrass M-test, $\sum \frac{\sin nx}{n^2}$ is uniformly convergent in $\mathbf{R}$.

Since

$$\left| \frac{\sin nx}{n} \right| \le \frac{1}{n}, \quad \forall \, x \in \mathbf{R},$$

We take $M_n = \frac{1}{n}$. Then, by the p-test, the series $\sum \frac{1}{n}$ is divergent because $p = 1$. Therefore, we cannot apply the Weierstrass M-test. The series $\sum \frac{\sin nx}{n}$ may or may not be uniformly convergent.

Abel's Lemma. *Let $k_1 < \sum_{r=m}^{p} u_r < k_2$ for $p = m, m+1, \dots, n$ and let $\langle v_n \rangle$ be a monotonic decreasing sequence of positive numbers. Then*

$$k_1 v_m < \sum_{r=m}^{n} u_r < k_2 v_m.$$

Theorem 2 (Abel's theorem). *Suppose that*

a. *$\sum u_n(x)$ is uniformly convergent on $[a, b]$.*
b. *The sequence $\langle v_n(x) \rangle$ is uniformly bounded in $[a, b]$. That is, there exists a positive number M such that $|v_n(x)| \le M, \forall x \in [a, b]$, and $\forall n \in \mathbf{N}$.*
c. *The sequence $\langle v_n(x) \rangle$ is monotonic decreasing in $[a, b]$.*

Then the series $\sum u_n(x) v_n(x)$ is uniformly convergent on $[a, b]$.

Proof. Since the $\sum u_n(x)$ is uniformly convergent on $[a, b]$, by the Cauchy general principle of uniform convergence, for each $\varepsilon > 0$ and

$\forall x \in [a, b]$, there exists a natural number m depending on ε only such that

$$|u_{n+1}(x) + u_{n+2}(x) + \cdots + u_{n+p}(x)| < \varepsilon,$$
$$\forall\, n \geq m, \quad \forall\, x \in [a, b] \quad \forall\, p \geq 1.$$

That is,

$$\left| \sum_{r=n+1}^{r=n+p} u_r(x) \right| < \frac{\varepsilon}{M}, \quad \forall\, n \geq m, \quad \forall\, x \in [a, b] \quad \forall\, p \geq 1. \qquad (8.4)$$

From the given condition:

$$|v_n(x)| \leq M, \quad \forall\, x \in [a, b], \quad \text{and} \quad \forall\, n \in \mathbf{N}, \qquad (8.5)$$

since $\langle v_n(x) \rangle$ is a monotonic decreasing sequence in $[a, b]$, by Abel's lemma and Eq. (8.4), we have

$$\left| \sum_{r=n+1}^{r=n+p} u_r(x) v_r(x) \right| < \frac{\varepsilon}{M} |v_{n+1}(x)|, \quad \forall\, n \geq m, \quad \forall\, x \in [a, b] \,\forall\, p \geq 1.$$
$$(8.6)$$

Again, since $\langle v_n(x) \rangle$ is a monotonic decreasing and uniformly bounded sequence in $[a, b]$, $|v_{n+1}(x)| \leq |v_n(x)|$. From Eqs. (8.5) and (8.6), we can write

$$\left| \sum_{r=n+1}^{r=n+p} u_r(x) v_r(x) \right| < \frac{\varepsilon}{M} |v_n(x)| < \frac{\varepsilon}{M} M = \varepsilon, \quad \forall\, n \geq m,$$
$$\forall\, x \in [a, b] \quad \forall\, p \geq 1.$$

That is,

$$|u_{n+1}(x)v_{n+1}(x) + u_{n+2}(x)v_{n+2}(x) + \cdots + u_{n+p}(x)v_{n+p}(x)| < \varepsilon,$$
$$\forall\, n \geq m, \quad \forall\, x \in [a, b] \quad \forall\, p \geq 1.$$

Therefore, by the Cauchy general principle of uniform convergence, the series $\sum u_n(x) v_n(x)$ is uniformly convergent on $[a, b]$.

Example 1. Show that the series $\sum \frac{(-1)^{n-1}}{n^2}|x|^n$ is uniformly convergent in $[-1,1]$.

Solution. Let $u_n(x) = \frac{(-1)^{n-1}}{n^2}$ and $v_n(x) = |x|^n$.

The series $\sum \frac{(-1)^{n-1}}{n^2}$ is alternating and monotonic decreasing, and $\lim_{n\to\infty} u_n = 0$. So, by the Leibniz test, it is convergent. Since the given series is free from the point x, it is uniformly convergent on $[-1,1]$.

Also, $v_n(x) = |x|^n$ is a positive monotonic decreasing sequence in n for a fixed value of $x \in [-1,1]$. Further, since $v_n(x) \leq 1, \forall x \in [-1,1]$, the sequence $\langle v_n(x) \rangle$ is uniformly bounded in $[-1,1]$.

Therefore, by Abel's test, the series $\sum u_n(x)v_n(x)$, that is, $\sum \frac{(-1)^{n-1}}{n^2}|x|^n$, is uniformly convergent in $[-1,1]$.

Example 2. If $\sum a_n$ is a convergent series of positive constants, then show that the series $\sum \frac{a_n x^n}{1+x^n}$ is uniformly convergent in $[0,1]$.

Solution. Let $u_n(x) = a_n$ and $v_n(x) = \frac{x^n}{1+x^n}$.

Since $\sum a_n$ is a convergent series of positive constants, it will be independent from point x and hence it will be uniformly convergent in $[0,1]$.

Also $v_n(x) = \frac{x^n}{1+x^n}$ is a positive monotonic decreasing sequence in n for a fixed value of $x \in [0,1]$. Further, $v_n(x) \leq 1, \forall x \in [0,1]$. So the sequence $\langle v_n(x) \rangle$ is uniformly bounded in $[0,1]$.

Therefore, by Abel's test, the series $\sum u_n(x)v_n(x)$, that is, $\sum \frac{a_n x^n}{1+x^n}$, is uniformly convergent in $[0,1]$.

Theorem 3 (Dirichlet's theorem). *Suppose that*

a. *There exists a positive constant M such that $|S_n(x)| = |\sum_{r=1}^{n} u_r(x)| < M, \forall x \in [a,b]$, and $\forall n \in \mathbf{N}$.*
b. *The sequence $\langle v_n(x) \rangle$ is positive monotonic decreasing and uniformly convergent to zero in $[a,b]$.*

Then the series $\sum u_n(x)v_n(x)$ is uniformly convergent on $[a,b]$.

Proof.   Given

$$|S_n(x)| = \left|\sum_{r=1}^{n} u_r(x)\right| < M, \forall x \in [a,b], \quad \text{and} \quad \forall\, n \in \mathbf{N},$$

then

$$|S_{n+p}(x)| = \left| \sum_{r=1}^{n+p} u_r(x) \right| < M, \forall x \in [a,b], \quad \forall\, p \geq 1, \text{ and } \quad \forall\, n \in \mathbf{N}.$$

Now, we have

$$|S_{n+p}(x) - S_n(x)| \leq |S_{n+p}(x)| + |S_n(x)| < 2M.$$

$$\Rightarrow |S_{n+p}(x) - S_n(x)| = \left| \sum_{r=1}^{n+p} u_r(x) - \sum_{r=1}^{n} u_r(x) \right|$$

$$= \left| \sum_{r=n+1}^{n+p} u_r(x) \right| < 2M.$$

$$\Rightarrow \quad \left| \sum_{r=n+1}^{n+p} u_r(x) \right| < 2M, \quad \forall\, x \in [a,b], \quad \forall\, p \geq 1, \text{ and } \forall\, n \geq m.$$

$$(8.7)$$

Since $\langle v_n(x) \rangle$ is a positive monotonic decreasing sequence in $[a,b]$, by using Abel's lemma in Eq. (8.7), we get

$$\Rightarrow \left| \sum_{r=n+1}^{n+p} u_r(x) v_r(x) \right| < 2M v_{n+1}(x), \quad \forall\, x \in [a,b],$$

$$\forall\, p \geq 1, \text{ and } \forall\, n \geq m_1. \qquad (8.8)$$

Since the sequence $\langle v_n(x) \rangle$ is uniformly convergent to zero in $[a,b]$, for each $\varepsilon > 0$, there exists a natural number m_2 such that

$$|v_n(x) - 0| = |v_n(x)| = v_n(x) \leq \frac{\varepsilon}{2M}, \quad \forall\, n \geq m_2. \qquad (8.9)$$

Let $m = \max\{m_1, m_2\}$. Then Eqs. (8.8) and (8.9) can be written as follows:

$$\left| \sum_{r=n+1}^{n+p} u_r(x) v_r(x) \right|$$

$$< 2M v_{n+1}(x), \quad \forall\, x \in [a,b], \quad \forall\, p \geq 1, \text{ and } \forall\, n \geq m. \quad (8.10)$$

$$v_n(x) \le \frac{\varepsilon}{2M}, \quad \forall n \ge m. \tag{8.11}$$

From Eqs. (8.10) and (8.11), we get

$$\left| \sum_{r=n+1}^{n+p} u_r(x) v_r(x) \right| < 2M v_{n+1}(x) < 2M v_n(x)$$

$$< 2M \frac{\varepsilon}{2M} = \varepsilon, \quad \forall\, x \in [a, b], \quad \forall\, p \ge 1, \text{ and } \forall\, n \ge m.$$

$$\Rightarrow \left| \sum_{r=n+1}^{n+p} u_r(x) v_r(x) \right| < \varepsilon, \quad \forall\, x \in [a, b], \quad \forall\, p \ge 1, \text{ and } \forall n \ge m.$$

That is,

$$|u_{n+1}(x) v_{n+1}(x) + u_{n+2}(x) v_{n+2}(x) + \cdots + u_{n+p}(x) v_{n+p}(x)| < \varepsilon,$$

$$\forall\, n \ge m, \quad \forall\, x \in [a, b] \quad \forall\, p \ge 1.$$

Therefore, by the Cauchy general principle of uniform convergence, the series $\sum u_n(x) v_n(x)$ is uniformly convergent on $[a, b]$.

Example 1. Show that the series $\sum (-1)^n \frac{x^2+n}{n^2}$ is uniformly convergent in $[a, b]$.

Solution. Let $u_n(x) = (-1)^n$ and $v_n(x) = \frac{x^2+n}{n^2}$. Then

$$S_n(x) = \sum_{r=1}^{n} u_r(x) = \begin{cases} 0, & \text{if } n \text{ is even} \\ -1, & \text{if } n \text{ is odd.} \end{cases}$$

$$\Rightarrow |S_n(x)| \le 1.$$

There exists some constant M such that $|x| < M \quad \forall x \in [a, b]$.
   That is,

$$v_n(x) = \frac{x^2+n}{n^2} < \frac{M^2+n}{n^2}.$$

$$\Rightarrow \lim_{n \to \infty} v_n(x) = 0 \quad \forall\, x \in [a, b].$$

Clearly, the sequence $\langle v_n(x) \rangle$ is positive monotonic decreasing and uniformly convergent to zero in $[a, b]$.

By Dirichlet's test, the series $\sum u_n(x)v_n(x)$, that is, $\sum(-1)^n \frac{x^2+n}{n^2}$, is uniformly convergent on $[a, b]$.

Example 2. Show that the series $\sum \frac{\sin nx}{n^p}$, $p > 0$, is uniformly convergent in $[\beta, 2\pi - \beta]$, where $0 < \beta < \pi$.

Solution. When $p > 1$, then $\left|\frac{\sin nx}{n^p}\right| \leq \frac{1}{n^p}$, $\forall x \in [\beta, 2\pi - \beta]$, where $0 < \beta < \pi$.

Taking $M_n = \frac{1}{n^p}$, by the p-test, the series $\sum \frac{1}{n^p}$ is convergent for $p > 1$. Therefore, by Weierstrass's M-test, $\sum \frac{\sin nx}{n^p}$ is uniformly convergent in $[\beta, 2\pi - \beta]$, where $0 < \beta < \pi$ for $p > 1$.

When $0 < p \leq 1$, let $u_n(x) = \sin nx$ and $v_n(x) = \frac{1}{n^p}$. Then

$$S_n(x) = \sum_{r=1}^{n} \sin rx = \frac{\sin\left\{\frac{(n+1)}{2}x\right\}\sin\left(\frac{nx}{2}\right)}{\sin\left(\frac{x}{2}\right)}.$$

$$\Rightarrow |S_n(x)| = \left|\frac{\sin\left\{\frac{(n+1)}{2}x\right\}\sin\left(\frac{nx}{2}\right)}{\sin\left(\frac{x}{2}\right)}\right|$$

$$\leq \operatorname{cosec}\left(\frac{\beta}{2}\right), \quad \forall\, x \in [\beta, 2\pi - \beta], \quad \text{where } 0 < \beta < \pi.$$

$$\Rightarrow \lim_{n\to\infty} v_n(x) = 0, \quad \forall\, x \in [\beta, 2\pi - \beta], \text{ and } 0 < p \leq 1.$$

Clearly, the sequence $\langle v_n(x)\rangle$ is positive monotonic decreasing and uniformly convergent to zero in $[\beta, 2\pi - \beta]$.

By Dirichlet's test the series $\sum u_n(x)v_n(x)$, that is, $\sum \frac{\sin nx}{n^p}$, is uniformly convergent on $[\beta, 2\pi - \beta]$ for $0 < p \leq 1$.

So by combining Weierstrass M-test and Dirichlet's test, we conclude that series $\sum \frac{\sin nx}{n^p}$ is uniformly convergent for $p > 0$ on $[\beta, 2\pi - \beta]$, where $0 < \beta < \pi$.

Exercises

1. Examine the uniform convergence of the sequence $\langle f_n \rangle$ of functions defined by $f_n(x) = \tan^{-1} nx$, $x \geq 0$ on the interval $[a, b], a > 0$, and $[0, b]$.

Ans. Uniform convergent in $[a, b], a > 0$, but only point-wise on $[0, b]$.

2. Show that the sequence $\left\{ \frac{nx}{1+n^3x^2} \right\}$ converges uniformly on the interval $[0, 1]$.

3. Show that the sequence $\left\{ \frac{n^2x}{1+n^3x^2} \right\}$ is not uniformly convergent on the interval $[0, 1]$.

4. Show that the sequence $\left\{ \frac{x}{1+nx^2} \right\}$ converges uniformly on any closed interval.

5. Using the Weierstrass M-test, show that the following series converges uniformly for all real values of θ:

$$\sum k^n \cos n\theta, \ \sum k^n \sin n\theta, \ \sum k^n \cos n^2\theta,$$

$$\sum k^n \sin (b^n\theta), \quad 0 < k < 1.$$

Take $M_n = k^n$.

6. Show that the following series converges uniformly for all real values of x:

$$\sum \frac{\sin (x^2 + n^2x)}{n(n+1)}.$$

Take $M_n = \frac{1}{n(n+1)}$.

7. If $\sum a_n$ is a convergent series of positive constants, then show that the series $\sum \frac{a_n x^{2n}}{1+x^{2n}}$ is uniformly convergent in $[0, 1]$.

8. If $\sum a_n$ is a convergent series of positive constants, then show that the series $\sum \frac{a_n x^n}{1+x^{2n}}$ is uniformly convergent in $[0, 1]$.

9. Show that $\sum \frac{(-1)^n}{x^2+n^2}$ is uniformly convergent on **R**.

Chapter 9

Functions of Several Variables

In Chapter 6, we have discussed the limit and continuity of real functions of one variable. In this chapter, we extend our notion of limits and continuity of functions from one variable to functions of two variables. Firstly, we write some definitions and then define the limits and continuity of functions of two variables. Further, we read Taylor's theorem for the functions of two variables. In the next section, we give a basic idea for the maxima and minima of functions of two variables. In the last section, we use Lagrange's method of undetermined multipliers for finding the maxima and minima of functions of three variables.

9.1 Some Basic Definitions

Cartesian product. The Cartesian product of the set of real numbers is denoted as $\mathbf{R}^2$ and defined as

$$R^2 = \{(x, y) : x \in \mathbf{R} \text{ and } y \in \mathbf{R}\}.$$

Explicit function of two variables. Let z be a dependent variable and x, y be the independent variables such that

$$z = f(x, y).$$

That is, when the values of x and y are known, then we can calculate the value of z explicitly. These types of functions are called explicit functions of two variables.

Example 1. Let $f(x, y) = x + y$, where $\{(x, y) : 0 \leq x \leq 1$ and $0 \leq y \leq 1\}$. Then the function $f(x, y)$ is an explicit function of two variables.

Neighborhood of a point. The rectangular neighborhood of a point $(a, b) \in \mathbf{R}^2$ is defined as

$$N = \{(x, y) : |x - a| < \delta \quad \text{and} \quad |x - b| < \delta\} - (a, b) \text{ or}$$

$$\{(x, y) : a - \delta < x < a + \delta \quad \text{and} \quad b - \delta < y < b + \delta\} - (a, b).$$

Since we removed the points in (a, b) from the rectangle, we can also treat this neighborhood as a deleted neighborhood. In this chapter, a deleted neighborhood will be treated as a neighborhood.

9.2   Limits of Functions of Two Variables

A number l is said to be the limit of a function $f(x, y)$ of two variables at a point (a, b) if $f(x, y)$ tends to l as (x, y) tends to the point (a, b). That is, for each $\varepsilon > 0$, there exists a $\delta > 0$ such that

$$|f(x, y) - l| < \varepsilon \quad \text{whenever} \quad |x - a| < \delta \quad \text{and} \quad |y - b| < \delta$$

$$\text{and} \quad (xy) \neq (a, b).$$

$$\text{Or } |f(x, y) - l| < \varepsilon \quad \text{whenever} \quad (x, y) \in N.$$

Remark. In $\mathbf{R}$, there are only two possibilities for approaching the point at which limits are to be calculated. We calculate these two possibilities as the left-hand limit and the right-hand limit. If these two limits are equal, then the limit of the function exists at that point. But, in the case of $\mathbf{R}^2$, there are an infinite number of possibilities to approach the point at which the limit is to be calculated. It is not possible to calculate limits along each path. But, if we find two different paths along which the limits are distinct, then we can say that the limit of the function of two variables will not exist. So it is easy to show the non-existence of the limit of functions of two variables. To show the existence of a limit, we will move to the following definition.

Non-existence of limits. If $\lim_{(x,y)\to(a,b)} f(x, y) = l$, then the limit will be l if we approach the point (a, b) along any curve. Let

$y = \theta(x)$ be any curve such that $\theta(x) \to b$, when $x \to a$. Then $\lim_{x \to a} f(x, \theta(x))$ must exist and should be equal to l.

Thus, if we can find two such curves $\theta_1(x)$ and $\theta_2(x)$ such that $\lim_{x \to a} f(x, \theta_1(x))$ and $\lim_{x \to a} f(x, \theta_2(x))$ are different, then the limit of the function $f(x, y)$ will not exist at the point (a, b).

Or if $\lim_{(x,y) \to (a,b)} f(x, y) = l$ and if $x = \theta(y)$ is any curve such that $\theta(y) \to a$, when $y \to b$, then $\lim_{y \to b} f(\theta(y), y)$ must exist and should be equal to l. Thus, if we can find two curves $\theta_1(y)$ and $\theta_2(y)$ such that $\lim_{y \to b} f(\theta_1(y), y)$ and $\lim_{y \to b} f(\theta_2(y), y)$ are different, then the limit of the function $f(x, y)$ will not exist at point (a, b).

Example 1. Find $\lim_{(x,y) \to (0,0)} f(x, y)$, where $f(x, y) = \frac{2xy}{x^2+y^2}$.

Solution. If we approach point $(0, 0)$ along the curve $y = mx$, then we get

$$\lim_{(x,y) \to (0,0)} f(x, y) = \lim_{x \to 0} f(x, mx) = \lim_{x \to 0} \frac{2mx^2}{x^2 + (mx)^2}$$

$$= \lim_{x \to 0} \frac{2m}{1 + m^2} = \frac{2m}{1 + m^2}.$$

If we take $m = 1$, that is, the curve $y = x$, then $\lim_{(x,y) \to (0,0)} f(x, y) = 1$.

If we take $m = 2$, that is, the curve $y = 2x$, then $\lim_{(x,y) \to (0,0)} f(x, y) = \frac{4}{5}$.

Since the limits are different along these different curves, therefore $\lim_{(x,y) \to (0,0)} \frac{2xy}{x^2+y^2}$ does not exist.

Example 2. Find $\lim_{(x,y) \to (0,0)} f(x, y)$, where $f(x, y) = \frac{xy^3}{x^2+y^6}$.

Solution. If we approach the point $(0, 0)$ along the curve $x = my^3$, then we get

$$\lim_{(x,y) \to (0,0)} f(x, y) = \lim_{y \to 0} f(my^3, y) = \lim_{x \to 0} \frac{my^6}{(my^3)^2 + y^6}$$

$$= \lim_{x \to 0} \frac{m}{1 + m^2} = \frac{m}{1 + m^2}.$$

If we take $m = 1$, that is, the curve $x = y^3$, then $\lim_{(x,y) \to (0,0)} f(x, y) = \frac{1}{2}$.

If we take $m = 2$, that is, the curve $x = 2y^3$, then $\lim_{(x,y) \to (0,0)}$ $f(x, y) = \frac{2}{5}$.

Since the limits are different along the different curves, $\lim_{(x,y) \to (0,0)} \frac{xy^3}{x^2 + y^6}$ does not exist.

Example 3. Find $\lim_{(x,y) \to (0,0)} f(x, y)$, where $f(x, y) = \frac{x^2 + y^2}{x - y}$.

Solution. If we approach the point $(0, 0)$ along the curve $y = x - mx^2$, then we get

$$\lim_{(x,y) \to (0,0)} f(x, y) = \lim_{x \to 0} f(x, x - mx^2) = \lim_{x \to 0} \frac{x^2 + (x - mx^2)^2}{x - (x - mx^2)}$$

$$= \lim_{x \to 0} \frac{2 + m^2 x^2 - 2mx}{m} = \frac{2}{m}.$$

If we take $m = 1$, that is, the curve $y = x$, then $\lim_{(x,y) \to (0,0)}$ $f(x, y) = 2$.

If we take $m = 2$, that is, the curve $y = 2x$, then $\lim_{(x,y) \to (0,0)}$ $f(x, y) = 1$.

Since the limits are different along these different curves, $\lim_{(x,y) \to (0,0)} \frac{x^2 + y^2}{x - y}$ does not exist.

Example 4. Find $\lim_{(x,y) \to (0,0)} f(x, y)$, where $f(x, y) = \frac{x^2 y^2}{x^2 y^2 + (x^2 - y^2)^2}$.

Solution. If we approach the point $(0, 0)$ along the curve $y = mx$, then we get

$$\lim_{(x,y) \to (0,0)} f(x, y) = \lim_{x \to 0} f(x, mx) = \lim_{x \to 0} \frac{x^2 (mx)^2}{x^2 (mx)^2 + (x^2 - (mx)^2)^2}$$

$$= \frac{m^2}{m^2 + (1 - m^2)^2}.$$

If we take $m = 1$, that is, the curve $y = x$, then $\lim_{(x,y) \to (0,0)}$ $f(x, y) = 1$.

If we take $m = 2$, that is, the curve $y = 2x$, then $\lim_{(x,y) \to (0,0)}$ $f(x, y) = \frac{4}{13}$.

Since the limits are different along these different curves, $\lim_{(x,y)\to(0,0)} \dfrac{x^2y^2}{x^2y^2+(x^2-y^2)^2}$ does not exist.

Example 5. Using ε, δ definition, show that $\lim_{(x,y)\to(2,3)} (4x + 2y) = 14$.

Solution. Here $f(x, y) = 4x + 2y$. Now,

$$
\begin{aligned}
|f(x, y) - 14| &= |4x + 2y - 14| = |4x - 8 + 2y - 6| \\
&= |4(x - 2) + 2(y - 3)| \\
&\le 4|x - 2| + 2|y - 3| \\
|f(x, y) - 14| &\le 4|x - 2| + 2|y - 3|.
\end{aligned}
\tag{9.1}
$$

Let $\varepsilon > 0$ and choose $\delta = \frac{\varepsilon}{8}$. Then, for $|x - 2| < \delta$ and $|y - 3| < \delta$, from Eq. (9.1), we have

$$
\begin{aligned}
|f(x, y) - 14| &\le 4\frac{\varepsilon}{8} + 2\frac{\varepsilon}{8} = \frac{\varepsilon}{2} + \frac{\varepsilon}{4} < \frac{\varepsilon}{2} + \frac{\varepsilon}{2} = \varepsilon. \\
|f(x, y) - 14| &< \varepsilon
\end{aligned}
$$

whenever $|x - 2| < \delta$ and $|y - 3| < \delta$.
Therefore, $\lim_{(x,y)\to(2,3)} (4x + 2y) = 14$.

Example 6. Using the ε, δ definition, show that $\lim_{(x,y)\to(1,2)} 3xy = 6$.

Solution. Here $f(x, y) = 3xy$. Now,

$$
\begin{aligned}
|f(x, y) - 6| &= |3xy - 6| = |3xy - 3y + 3y - 6| \\
&= |3y(x - 1) + 3(y - 2)| \\
&\le 3|y||x - 1| + 3|y - 2| \\
|f(x, y) - 6| &\le 3|y||x - 1| + 3|y - 2|.
\end{aligned}
\tag{9.2}
$$

Let $|y - 2| < \delta_1$ and take $\delta_1 = 1$. Then $|y - 2| < 1$.

$$
\Rightarrow |y| \le 3.
\tag{9.3}
$$

From Eqs. (9.2) and (9.3), for $|y - 2| < \delta_1$, we have

$$|f(x, y) - 6| \leq 9|x - 1| + 3|y - 2|. \tag{9.4}$$

Let $\varepsilon > 0$ and choose $\delta = \min\{1, \frac{\varepsilon}{18}\}$. Then, for $|x - 1| < \delta$ and $|y - 2| < \delta$, from Eq. (9.4), we have

$$|f(x, y) - 6| \leq 9\frac{\varepsilon}{18} + 3\frac{\varepsilon}{18} = \frac{\varepsilon}{2} + \frac{\varepsilon}{6} < \frac{\varepsilon}{2} + \frac{\varepsilon}{2} = \varepsilon.$$

That is, $|f(x, y) - 6| < \varepsilon$ whenever $|x - 1| < \delta$ and $|y - 2| < \delta$. Therefore, $\lim_{(x,y)\to(1,2)} 3xy = 6$.

Example 7. Using the ε, δ definition, show that $\lim_{(x,y)\to(0,0)}$ $\frac{x^3-y^3}{x^2+y^2} = 0$.

Solution. Here $f(x, y) = \frac{x^3-y^3}{x^2+y^2}$. Now,

$$|f(x, y) - 0| = \left|\frac{x^3 - y^3}{x^2 + y^2} - 0\right| = \left|\frac{x^3 - y^3}{x^2 + y^2}\right|.$$

Put $x = r\cos\theta$ and $y = r\sin\theta$, so that

$$\left|\frac{x^3 - y^3}{x^2 + y^2}\right| = |r(\cos^3\theta - \sin^3\theta)| \leq r(|\cos^3\theta| + |\sin^3\theta|) \leq 2r$$

$$= 2\sqrt{x^2 + y^2} < \varepsilon, \tag{9.5}$$

if

$$4x^2 < \frac{\varepsilon^2}{2} \quad \text{and} \quad 4y^2 < \frac{\varepsilon^2}{2}$$

or if

$$|x| < \frac{\varepsilon}{2\sqrt{2}} \quad \text{and} \quad |y| < \frac{\varepsilon}{2\sqrt{2}}.$$

Let $\varepsilon > 0$ and choose $\delta = \frac{\varepsilon}{2\sqrt{2}}$. Then, for $|x - 0| < \delta$ and $|y - 0| < \delta$, from Eq. (9.5), we have

$$\left|\frac{x^3 - y^3}{x^2 + y^2} - 0\right| < \varepsilon.$$

Therefore, $\lim_{(x,y)\to(0,0)} \frac{x^3-y^3}{x^2+y^2} = 0$.

Example 8. Using the ε, δ definition, show that $\lim_{(x,y)\to(0,0)}$ $\frac{x^3 y^3}{x^2 + y^2} = 0$.

Solution. Here $f(x, y) = \frac{x^3 y^3}{x^2 + y^2}$. Now,

$$|f(x, y) - 0| = \left| \frac{x^3 y^3}{x^2 + y^2} - 0 \right| = \left| \frac{x^3 y^3}{x^2 + y^2} \right|.$$

Put $x = r \cos \theta$ and $y = r \sin \theta$, then we get

$$\left| \frac{x^3 y^3}{x^2 + y^2} \right| = |r^4 \cos^3 \theta \sin^3 \theta| = r^4 |\cos^3 \theta \sin^3 \theta| \le r^4$$

$$= (x^2 + y^2)^2 < \varepsilon, \tag{9.6}$$

if

$$x^2 < \frac{\sqrt{\varepsilon}}{2} \quad \text{and} \quad y^2 < \frac{\sqrt{\varepsilon}}{2}$$

or if

$$|x| < \left(\frac{\sqrt{\varepsilon}}{2} \right)^{\frac{1}{2}} \quad \text{and} \quad |y| < \left(\frac{\sqrt{\varepsilon}}{2} \right)^{\frac{1}{2}}.$$

Let $\varepsilon > 0$ and choose $\delta = (\frac{\sqrt{\varepsilon}}{2})^{\frac{1}{2}}$. Then, for $|x - 0| < \delta$ and $|y - 0| < \delta$, from Eq. (9.6), we have

$$\left| \frac{x^3 y^3}{x^2 + y^2} - 0 \right| < \varepsilon.$$

Therefore, $\lim_{(x,y)\to(0,0)} \frac{x^3 y^3}{x^2 + y^2} = 0$.

9.3 Repeated Limits

If a function f is defined in some neighborhood of (a, b), then $\lim_{x\to a} f(x, y)$, if it exists, is a function of y, say $\theta(y)$. Then, if the limit $\lim_{y\to b} \theta(y)$ exists and is equal to λ, we write

$$\lim_{y\to b} \lim_{x\to a} f(x, y) = \lambda.$$

Then λ is known as a repeated limit of the function f as $x \to a$, $y \to b$.

If we change the order of the limit by taking first $y \to b$, then $x \to a$, we get the other repeated limit which is denoted by λ' and given as follows:

$$\lim_{x \to a} \lim_{y \to b} f(x, y) = \lambda'.$$

These two limits may or may not be equal.

Example 1. Calculate the repeated limits and the simultaneous limit for the function $f(x, y) = \frac{3xy}{x^2+y^2}$ at the origin.

Solution. Let $(x, y) = \frac{3xy}{x^2+y^2}$. Then

$$\lim_{y \to 0} \lim_{x \to 0} f(x, y) = (0) \lim_{y \to 0} = 0,$$

$$\lim_{x \to 0} \lim_{y \to 0} f(x, y) = (0) \lim_{x \to 0} = 0.$$

Thus, the repeated limits exist and are equal.

For the simultaneous limit, if we approach the point $(0, 0)$ along the curve $y = mx$, then we get

$$\lim_{(x,y) \to (0,0)} f(x, y) = \lim_{x \to 0} f(x, mx) = \lim_{x \to 0} \frac{3mx^2}{x^2 + (mx)^2}$$

$$= \lim_{x \to 0} \frac{3m}{1 + m^2} = \frac{3m}{1 + m^2}.$$

If we take $m = 1$, that is, the curve $y = x$, then $\lim_{(x,y) \to (0,0)} f(x, y) = \frac{3}{2}$.

If we take $m = 2$, that is, the curve $y = 2x$, then $\lim_{(x,y) \to (0,0)} f(x, y) = \frac{6}{5}$.

Since the limits are different along these different curves, therefore, the simultaneous limit $\lim_{(x,y) \to (0,0)} \frac{3xy}{x^2+y^2}$ does not exist.

Remark. If the repeated limits are not equal, then the simultaneous limit does not exist.

Example 2. Calculate the repeated limits and the simultaneous limit for the function $f(x, y) = \frac{y-x}{y+x} \frac{1+x}{1+y}$ at the origin.

Solution. Let $(x, y) = \frac{y-x}{y+x} \frac{1+x}{1+y}$. Then

$$\lim_{y \to 0} \lim_{x \to 0} f(x, y) = \lim_{y \to 0} \left(\frac{1}{1+y} \right) = 1$$

and

$$\lim_{x \to 0} \lim_{y \to 0} f(x, y) = \lim_{x \to 0} \left(-\frac{1+x}{1} \right) = -1.$$

Thus, both of the repeated limits exist, but are not equal. From the above remark, we can conclude that the simultaneous limit does not exist, which can be verified by taking $y = mx$.

9.4 Continuity of Functions of Two Variables

A function f is said to be continuous at a point (a, b) of its domain if the limit of the function exists at that point and it is equal to the value of the function at that point. That is,

$$\lim_{(x,y) \to (a,b)} f(x, y) = f(a, b).$$

Alternatively, a function f is said to be continuous at a point (a, b) of its domain if, for each $\varepsilon > 0$, there exists a neighborhood N of (a, b) such that

$$|f(x, y) - l| < \varepsilon \quad \text{whenever} \quad (xy) \in N.$$

Or

$$|f(x, y) - f(a, b)| < \varepsilon \quad \text{whenever} \quad |x - a| < \delta \quad \text{and} \quad |x - b| < \delta.$$

Example 1. Discuss the continuity of the function $f(x, y)$ at $(x, y) = (0, 0)$, where

$$f(x, y) = \begin{cases} \dfrac{2xy}{x^2 + y^2}, & (x, y) \neq (0, 0), \\ 0, & (x, y) = (0, 0). \end{cases}$$

Solution. First, we check the limit of the function if we approach the point $(0,0)$ along the curve $y = mx$. Then we get

$$\lim_{(x,y)\to(0,0)} f(x,y) = \lim_{x\to0} f(x,mx) = \lim_{x\to0} \frac{2mx^2}{x^2 + (mx)^2}$$

$$= \lim_{x\to0} \frac{2m}{1+m^2} = \frac{2m}{1+m^2}.$$

If we take $m = 1$, that is, the curve $y = x$, then $\lim_{(x,y)\to(0,0)} f(x,y) = 1$.

If we take $m = 2$, that is, the curve $y = 2x$, then $\lim_{(x,y)\to(0,0)} f(x,y) = \frac{4}{5}$.

Since, the limits are different along these different curves, so $\lim_{(x,y)\to(0,0)} \frac{2xy}{x^2+y^2}$ does not exist.

Since the limit of the function does not exist, it is not continuous at the origin.

Example 2. Discuss the continuity of the function $f(x,y)$ at $(x,y) = (0,0)$ where

$$f(x,y) = \begin{cases} \dfrac{x^2y^2}{x^2 + y^2}, & (x,y) \neq (0,0), \\ 0, & (x,y) = (0,0). \end{cases}$$

Solution. First, we check the limit of the function $f(x,y) = \frac{x^2y^2}{x^2+y^2}$ at $(0,0)$. Now,

$$|f(x,y) - 0| = \left| \frac{x^2y^2}{x^2 + y^2} - 0 \right| = \left| \frac{x^2y^2}{x^2 + y^2} \right|.$$

If we put $x = r\cos\theta$ and $y = r\sin\theta$, then we get

$$\left| \frac{x^2y^2}{x^2 + y^2} \right| = |r^2 \cos^2\theta \sin^2\theta| = r^2|\cos^2\theta \sin^2\theta| \leq r^2$$

$$= x^2 + y^2 < \varepsilon, \tag{9.7}$$

if

$$x^2 < \frac{\varepsilon}{2} \quad \text{and} \quad y^2 < \frac{\varepsilon}{2},$$

or if

$$|x| < \sqrt{\frac{\varepsilon}{2}} \quad \text{and} \quad |y| < \sqrt{\frac{\varepsilon}{2}}.$$

Let $\varepsilon > 0$ and choose $\delta = \sqrt{\frac{\varepsilon}{2}}$. Then, for $|x - 0| < \delta$ and $|y - 0| < \delta$, from Eq. (9.7), we have

$$\left| \frac{x^2 y^2}{x^2 + y^2} - 0 \right| < \varepsilon.$$

Therefore, $\lim_{(x,y) \to (0,0)} \frac{x^2 y^2}{x^2 + y^2} = 0$.

Since $\lim_{(x,y) \to (0,0)} \frac{x^2 y^2}{x^2 + y^2} = f(0,0) = 0$, therefore, $f(x,y)$ is continuous at the origin.

Example 3. Discuss the continuity of the function $f(x,y)$ at $(2,3)$, where

$$f(x,y) = \begin{cases} x + y, & (x,y) \neq (2,3), \\ 5, & (x,y) = (2,3). \end{cases}$$

Solution. Here $f(x,y) = x + y$. Now,

$$|f(x,y) - 5| = |x + y - 5| = |x - 2 + y - 3| = |(x - 2) + (y - 3)|$$
$$\leq |x - 2| + |y - 3|$$
$$|f(x,y) - 5| \leq |x - 2| + |y - 3|. \tag{9.8}$$

Let $\varepsilon > 0$ and choose $\delta = \frac{\varepsilon}{2}$. Then, for $|x - 2| < \delta$ and $|y - 3| < \delta$, from Eq. (9.8), we have

$$|f(x,y) - 5| < \frac{\varepsilon}{2} + \frac{\varepsilon}{2} = \varepsilon.$$
$$|f(x,y) - 5| < \varepsilon$$

whenever $|x - 2| < \delta$ and $|y - 3| < \delta$.

Therefore, $\lim_{(x,y) \to (2,3)} x + y = 5$.

Since $\lim_{(x,y) \to (2,3)} x + y = f(2,3) = 5$, therefore, $f(x,y)$ is continuous at $(2,3)$.

Example 4. Discuss the continuity of the function $f(x, y)$ at $(1, 2)$ where

$$f(x, y) = \begin{cases} 2xy, & (x, y) \neq (1, 2), \\ 4, & (x, y) = (0, 0). \end{cases}$$

Solution. Here $f(x, y) = 2xy$. Now,

$$|f(x, y) - 4| = |2xy - 4| = |2xy - 2y + 2y - 4|$$
$$= |2y(x - 1) + 2(y - 2)|$$
$$\leq 2|y||x - 1| + 2|y - 2|$$
$$|f(x, y) - 4| \leq 2|y||x - 1| + 2|y - 2|. \tag{9.9}$$

Let $|y - 2| < \delta_1$, taking $\delta_1 = 1$, then $|y - 2| < 1$.

$$\Rightarrow |y| \leq 3. \tag{9.10}$$

From Eqs. (9.9) and (9.10), for $|y - 2| < \delta_1$, we have

$$|f(x, y) - 4| \leq 6|x - 1| + 2|y - 2|. \tag{9.11}$$

Let $\varepsilon > 0$ and choose $\delta = \min\{1, \frac{\varepsilon}{12}\}$. Then, for $|x - 1| < \delta$ and $|y - 2| < \delta$, from Eq. (9.11), we have

$$|f(x, y) - 4| \leq 6.\frac{\varepsilon}{12} + 2.\frac{\varepsilon}{12} = \frac{\varepsilon}{2} + \frac{\varepsilon}{6} < \frac{\varepsilon}{2} + \frac{\varepsilon}{2} = \varepsilon.$$
$$|f(x, y) - 4| < \varepsilon \quad \text{whenever} \quad |x - 1| < \delta \quad \text{and} \quad |y - 2| < \delta.$$

Therefore, $\lim_{(x,y)\to(1,2)} 2xy = 4$.

Since $\lim_{(x,y)\to(1,2)} 2xy = f(1, 2) = 4$, therefore, $f(x, y)$ is continuous at $(1, 2)$.

9.5   Taylor's Theorem for Functions of Two Variables

Statement. Let $f(x, y)$ be a function defined on a domain **D** whose partial derivatives up to the mth order exist and whose partial derivatives up to the $(m-1)$th order are continuous. If (c, d) and $(c+h, d+k) \in \mathbf{D}$, then

$$
\begin{aligned}
f(c+h, d+k) = f(c, d) &+ \left(h\frac{\partial}{\partial x} + k\frac{\partial}{\partial y} \right) f(c, d) \\
&+ \frac{1}{2!} \left(h\frac{\partial}{\partial x} + k\frac{\partial}{\partial y} \right)^2 f(c, d) + \cdots \\
&+ \frac{1}{(m-1)!} \left(h\frac{\partial}{\partial x} + k\frac{\partial}{\partial y} \right)^{m-1} f(c, d) + R_m.
\end{aligned}
$$

Here

$$
R_m = \frac{1}{m!} \left(h\frac{\partial}{\partial x} + k\frac{\partial}{\partial y} \right)^m f(c+\theta h, d+\theta k) \quad \text{and} \quad 0 < \theta < 1.
$$

If $c + h = x$ and $d + k = y$ then Taylor's expansion is written as follows:

$$
\begin{aligned}
f(x, y) = f(c, d) &+ \left((x-c)\frac{\partial}{\partial x} + (y-d)\frac{\partial}{\partial y} \right) f(c, d) \\
&+ \frac{1}{2!} \left((x-c)\frac{\partial}{\partial x} + (y-d)\frac{\partial}{\partial y} \right)^2 f(c, d) \\
&+ \cdots + \frac{1}{(m-1)!} \left((x-c)\frac{\partial}{\partial x} + (y-d)\frac{\partial}{\partial y} \right)^{m-1} f(c, d) + R_m.
\end{aligned}
$$

Here

$$
\begin{aligned}
R_m = \frac{1}{m!} &\left((x-c)\frac{\partial}{\partial x} + (y-d)\frac{\partial}{\partial y} \right)^m f(c+\theta(x-c), d \\
&+ \theta(y-d)) \quad \text{and} \quad 0 < \theta < 1.
\end{aligned}
$$

Maclaurin's expansion. If we take Taylor's expansion about the origin, then we will get Maclaurin's expansion. Taking $c = 0$ and

$d = 0$ in Taylor's expansion, we have

$$f(x,y) = f(0,0) + \left(x\frac{\partial}{\partial x} + y\frac{\partial}{\partial y} \right) f(0,0)$$

$$+ \frac{1}{2!} \left(x\frac{\partial}{\partial x} + y\frac{\partial}{\partial y} \right)^2 f(0,0)$$

$$+ \cdots + \frac{1}{(m-1)!} \left(x\frac{\partial}{\partial x} + y\frac{\partial}{\partial y} \right)^{m-1} f(0,0) + R_m.$$

Here

$$R_m = \frac{1}{m!} \left(x\frac{\partial}{\partial x} + y\frac{\partial}{\partial y} \right)^m f(\theta x, \theta y), \quad \text{and} \quad 0 < \theta < 1.$$

Example 1. If $f(x,y) = e^{x+y}$, then find Taylor's formula for $m = 3$ and at the point $(0,0)$.

Solution. The Taylor's expansion about point $c = 0$ and $d = 0$, and $m = 3$, is given as $f(x,y) = f(0,0) + (x\frac{\partial}{\partial x} + y\frac{\partial}{\partial y})f(0,0) + \frac{1}{2!}(x\frac{\partial}{\partial x} + y\frac{\partial}{\partial y})^2 f(0,0) + R_3$,
where

$$R_3 = \frac{1}{3!} \left(x\frac{\partial}{\partial x} + y\frac{\partial}{\partial y} \right)^3 f(\theta x, \theta y) \quad \text{and} \quad 0 < \theta < 1.$$

$$\Rightarrow f(x,y) = f(0,0) + \left(x\frac{\partial}{\partial x} + y\frac{\partial}{\partial y} \right) f(0,0)$$

$$+ \frac{1}{2!} \left(x^2\frac{\partial^2}{\partial x^2} + y^2\frac{\partial^2}{\partial y^2} + xy\frac{\partial^2}{\partial x\partial y} + yx\frac{\partial^2}{\partial y\partial x} \right) f(0,0) + R_3,$$

$$\tag{9.12}$$

where,

$$R_3 = \frac{1}{3!} \left(x^3\frac{\partial^3}{\partial x^3} + y^3\frac{\partial^3}{\partial y^3} + 3x^2 y\frac{\partial^3}{\partial x^2\partial y} + 3xy^2\frac{\partial^3}{\partial x\partial y^2} \right) f(\theta x, \theta y).$$

Since $f(x, y) = e^{x+y}$, then we have

$$\frac{\partial f}{\partial x} = e^{x+y}, \quad \frac{\partial f}{\partial y} = e^{x+y}, \quad \frac{\partial^2 f}{\partial x^2} = e^{x+y},$$

$$\frac{\partial^2 f}{\partial y^2} = e^{x+y}, \quad \frac{\partial^2 f}{\partial x \partial y} = e^{x+y},$$

$$\frac{\partial^2 f}{\partial y \partial x} = e^{x+y}, \quad \frac{\partial^3 f}{\partial x^3} = e^{x+y}, \quad \frac{\partial^3 f}{\partial y^3} = e^{x+y},$$

$$\frac{\partial^3 f}{\partial x^2 \partial y} = e^{x+y}, \quad \frac{\partial^3 f}{\partial x \partial y^2} = e^{x+y}.$$

Therefore, at point $(0, 0)$, we get

$$f(0,0) = 1, \quad \frac{\partial f}{\partial x} = 1, \quad \frac{\partial f}{\partial y} = 1, \quad \frac{\partial^2 f}{\partial x^2} = 1, \quad \frac{\partial^2 f}{\partial y^2} = 1, \quad \frac{\partial^2 f}{\partial x \partial y} = 1,$$

$$\frac{\partial^2 f}{\partial y \partial x} = 1.$$

At point $(\theta x, \theta y)$, we have

$$\frac{\partial^3 f}{\partial x^3} = \theta^3 e^{\theta(x+y)}, \quad \frac{\partial^3 f}{\partial y^3} = \theta^3 e^{\theta(x+y)}, \quad \frac{\partial^3 f}{\partial x^2 \partial y} = \theta^3 e^{\theta(x+y)},$$

$$\frac{\partial^3 f}{\partial x \partial y^2} = \theta^3 e^{\theta(x+y)}.$$

Using these values in Eq. (9.12), we get

$$e^{(x+y)} = 1 + (x + y) + \frac{1}{2!}(x^2 + y^2 + 2xy) + R_3,$$

where,

$$R_3 = \frac{\theta^3}{3!}(x^3 + y^3 + 3x^2 y + 3xy^2)e^{\theta(x+y)}, \quad \text{and} \quad 0 < \theta < 1.$$

Example 2. If $f(x, y) = e^x$, then find Taylor's formula for $m = 3$ and at point $(0, 0)$.

Solution. Taylor's expansion about the point $c = 0$ and $d = 0$, and for $m = 3$, is given as follows:

$$f(x, y) = f(0, 0) + \left(x \frac{\partial}{\partial x} + y \frac{\partial}{\partial y} \right) f(0, 0)$$

$$+ \frac{1}{2!} \left(x^2 \frac{\partial^2}{\partial x^2} + y^2 \frac{\partial^2}{\partial y^2} + xy \frac{\partial^2}{\partial x \partial y} + yx \frac{\partial^2}{\partial y \partial x} \right) f(0, 0) + R_3,$$

$$(9.13)$$

where

$$R_3 = \frac{1}{3!} \left(x^3 \frac{\partial^3}{\partial x^3} + y^3 \frac{\partial^3}{\partial y^3} + 3x^2 y \frac{\partial^3}{\partial x^2 \partial y} + 3xy^2 \frac{\partial^3}{\partial x \partial y^2} \right) f(\theta x, \theta y).$$

Since $f(x, y) = e^x$, we have

$$\frac{\partial f}{\partial x} = e^x, \quad \frac{\partial f}{\partial y} = 0, \quad \frac{\partial^2 f}{\partial x^2} = e^x, \quad \frac{\partial^2 f}{\partial y^2} = 0, \quad \frac{\partial^2 f}{\partial x \partial y} = 0,$$

$$\frac{\partial^2 f}{\partial y \partial x} = 0, \quad \frac{\partial^3 f}{\partial x^3} = e^x, \quad \frac{\partial^3 f}{\partial y^3} = 0, \quad \frac{\partial^3 f}{\partial x^2 \partial y} = 0, \quad \frac{\partial^3 f}{\partial x \partial y^2} = 0.$$

Therefore, at the point $(0, 0)$, we get

$$f(0, 0) = 1, \quad \frac{\partial f}{\partial x} = 1, \quad \frac{\partial f}{\partial y} = 0, \quad \frac{\partial^2 f}{\partial x^2} = 1, \quad \frac{\partial^2 f}{\partial y^2} = 0, \quad \frac{\partial^2 f}{\partial x \partial y} = 0,$$

$$\frac{\partial^2 f}{\partial y \partial x} = 0.$$

At the point $(\theta x, \theta y)$, we have

$$\frac{\partial^3 f}{\partial x^3} = \theta^3 e^{\theta x}, \quad \frac{\partial^3 f}{\partial y^3} = 0, \quad \frac{\partial^3 f}{\partial x^2 \partial y} = 0, \quad \frac{\partial^3 f}{\partial x \partial y^2} = 0.$$

Using these values in Eq. (9.13), we get

$$e^x = 1 + x + \frac{1}{2!} x^2 + R_3,$$

where

$$R_3 = \frac{\theta^3}{3!} x^3 e^{\theta x} \quad \text{and} \quad 0 < \theta < 1.$$

Example 3. If $f(x, y) = x^2 + xy - y^2$, then find Taylor's expansion in powers of $(x - 1)$ and $(y + 2)$.

Solution. We want to find Taylor's expansion in power of $(x-1)$ and $(y+2)$. Therefore, we consider $c = 1$ and $d = -2$ in Taylor's expansion. Since the given function is quadratic, so the third and higher derivatives will be zero. Thus, clearly, Taylor's expansion will be

$$f(x, y) = f(1, -2) + \left((x-1)\frac{\partial}{\partial x} + (y+2)\frac{\partial}{\partial y} \right) f(1, -2)$$

$$+ \frac{1}{2!} \left((x-1)\frac{\partial}{\partial x} + (y+2)\frac{\partial}{\partial y} \right)^2 f(1, -2).$$

$$f(x, y) = f(1, -2) + \left((x-1)\frac{\partial}{\partial x} + (y+2)\frac{\partial}{\partial y} \right) f(1, -2)$$

$$+ \frac{1}{2!} \left((x-1)^2\frac{\partial^2}{\partial x^2} + (y+2)^2\frac{\partial^2}{\partial y^2} + (x-1)(y+2)\frac{\partial^2}{\partial x \partial y} \right.$$

$$\left. + (y+2)(x-1)\frac{\partial^2}{\partial y \partial x} \right) f(1, -2).$$

Since $(x, y) = x^2 + xy - y^2$, then we have

$$\frac{\partial f}{\partial x} = 2x + y, \quad \frac{\partial f}{\partial y} = x - 2y, \quad \frac{\partial^2 f}{\partial x^2} = 2,$$

$$\frac{\partial^2 f}{\partial y^2} = -2, \quad \frac{\partial^2 f}{\partial x \partial y} = 1,$$

$$\frac{\partial^2 f}{\partial y \partial x} = 1, \quad \frac{\partial^3 f}{\partial x^3} = 0, \quad \frac{\partial^3 f}{\partial y^3} = 0, \quad \frac{\partial^3 f}{\partial x^2 \partial y} = 0, \quad \frac{\partial^3 f}{\partial x \partial y^2} = 0.$$

Therefore, at the point $(1, -2)$, we get

$$f(1, -2) = -5, \quad \frac{\partial f}{\partial x} = 0, \quad \frac{\partial f}{\partial y} = 5, \quad \frac{\partial^2 f}{\partial x^2} = 2,$$

$$\frac{\partial^2 f}{\partial y^2} = -2, \quad \frac{\partial^2 f}{\partial x \partial y} = 1,$$

$$\frac{\partial^2 f}{\partial y \partial x} = 1.$$

Therefore, we have

$$x^2 + xy - y^2 = -5 + 5(y + 2) + \frac{1}{2!}(2(x - 1)^2$$

$$- 2(y + 2)^2 + 2(x - 1)(y + 2)),$$

$$x^2 + xy - y^2 = -5 + 5(y + 2) + (x - 1)^2$$

$$- (y + 2)^2 + (x - 1)(y + 2).$$

Example 4. If $f(x, y) = \frac{y^2}{x^3}$, then find Taylor's expansion up to second-degree terms in powers of $(x - 1)$ and $(y - 1)$.

Solution. We want to find Taylor's expansion in powers of $(x - 1)$ and $(y - 1)$. Then we consider $c = 1$ and $d = 1$ in Taylor's expansion. Since we want to calculate Taylor's expansion up to the second-degree terms, Taylor's expansion will be

$$f(x, y) = f(1, 1) + \left((x - 1)\frac{\partial}{\partial x} + (y - 1)\frac{\partial}{\partial y} \right) f(1, 1)$$

$$+ \frac{1}{2!} \left((x - 1)\frac{\partial}{\partial x} + (y - 1)\frac{\partial}{\partial y} \right)^2 f(1, 1).$$

$$f(x, y) = f(1, 1) + \left((x - 1)\frac{\partial}{\partial x} + (y - 1)\frac{\partial}{\partial y} \right) f(1, 1)$$

$$+ \frac{1}{2!} \left((x - 1)^2\frac{\partial^2}{\partial x^2} + (y - 1)^2\frac{\partial^2}{\partial y^2} + (x - 1)(y - 1)\frac{\partial^2}{\partial x \partial y} \right.$$

$$\left. + (y - 1)(x - 1)\frac{\partial^2}{\partial y \partial x} \right) f(1, 1).$$

Since $(x, y) = \frac{y^2}{x^3}$, we have

$$\frac{\partial f}{\partial x} = -\frac{3y^2}{x^4}, \quad \frac{\partial f}{\partial y} = \frac{2y}{x^3}, \quad \frac{\partial^2 f}{\partial x^2} = \frac{12y^2}{x^5},$$

$$\frac{\partial^2 f}{\partial y^2} = \frac{2}{x^3}, \quad \frac{\partial^2 f}{\partial x \partial y} = -\frac{6y}{x^4},$$

$$\frac{\partial^2 f}{\partial y \partial x} = -\frac{6y}{x^4}.$$

Therefore, at the point $(1,1)$, we get

$$f(1,1) = 1, \quad \frac{\partial f}{\partial x} = -3, \quad \frac{\partial f}{\partial y} = 2, \quad \frac{\partial^2 f}{\partial x^2} = 12,$$

$$\frac{\partial^2 f}{\partial y^2} = 2, \quad \frac{\partial^2 f}{\partial x \partial y} = -6,$$

$$\frac{\partial^2 f}{\partial y \partial x} = -6.$$

Hence, we have

$$\frac{y^2}{x^3} = 1 - 3(x-1) + 2(y-1) + \frac{1}{2!}(12(x-1)^2$$

$$+ 2(y-1)^2 - 12(x-1)(y-1)),$$

$$\frac{y^2}{x^3} = 1 - 3(x-1) + 2(y-1) + 6(x-1)^2$$

$$+ (y-1)^2 - 6(x-1)(y-1).$$

9.6 Maxima and Minima of Functions of Two Variables

Let (a,b) be a point of domain of function f. Then $f(a,b)$ is an extreme value of function $f(x,y)$ if for every point (x,y) of some neighborhood of (a,b) the difference

$$f(x,y) - f(a,b) \tag{9.14}$$

keeps the same sign.

Depending on the sign of (9.14), positive or negative, the extreme values of $f(a,b)$ are known as the minimum and maximum values.

Necessary condition: A necessary condition for $f(x,y)$ to have an extreme value at (a,b) is that $f_x(a,b) = 0$ and $f_y(a,b) = 0$, provided that these partial derivatives exist.

Sufficient condition: Let $f_x(a,b) = 0$ and $f_y(a,b) = 0$. Suppose that $f(x,y)$ possesses the second-order partial derivatives in a given neighborhood of (a,b). Suppose that $f_{xx}(a,b)$, $f_{xy}(a,b)$ and $f_{yy}(a,b)$ are not all zero. Let $f(a+h, b+k)$ be a point of this neighborhood.

Let us define

$$r = f_{xx}(a, b), s = f_{xy}(a, b), t = f_{yy}(a, b).$$

By Taylor's theorem, we have, for $0 < \theta < 1$,

$$f(a + h, b + k) = f(a, b) + [hf_x(a, b) + kf_y(a, b)]$$
$$+ \frac{1}{2!}[h^2 f_{xx}(a + \theta h, b + \theta k)$$
$$+ 2hk f_{xy}(a + \theta h, b + \theta k) + k^2 f_{yy}(a + \theta h, b + \theta k)]$$

But the necessary condition for the extreme value is that $f_x(a, b) = 0$ and $f_y(a, b) = 0$.

Since the second-order partial derivatives are continuous at (a, b), we write

$$f_{xx}(a + \theta h, b + \theta k) - f_{xx}(a, b) = \alpha_1,$$
$$f_{xy}(a + \theta h, b + \theta k) - f_{xy}(a, b) = \alpha_2,$$
$$f_{yy}(a + \theta h, b + \theta k) - f_{yy}(a, b) = \alpha_3,$$

where $\alpha_1, \alpha_2, \alpha_3$ are functions of h and k, and tend to zero as $(h, k) \to (0, 0)$.

Therefore,

$$f(a + h, b + k) - f(a, b) = \frac{1}{2}[rh^2 + 2shk + tk^2 + \alpha],$$

where $\alpha = \alpha_1 + \alpha_2 + \alpha_3 \to 0$ as $(h, k) \to (0, 0)$ and is of an unknown sign.

Let

$$H = rh^2 + 2shk + tk^2.$$

Then the following cases can be considered.

Case 1. H keeps a constant sign and never vanishes.

Since $\alpha \to 0$ as $(h, k) \to (0, 0)$, therefore, α is a small number and the sign of $H + \alpha$ is the same as that of H, that is, $H + \alpha$ is positive or negative according as H is positive or negative. Thus, the difference

$$f(a + h, b + k) - f(a, b) > 0 \text{ if } H > 0 \quad \text{and}$$

$$f(a + h, b + k) - f(a, b) < 0 \text{ if } H < 0.$$

By definition, the function $f(x, y)$ has a minima or maxima at (a, b) according as the difference $f(a + h, b + k) - f(a, b)$ is positive or negative for all (h, k) except $(0, 0)$.

Thus, $f(a, b)$ will be the minimum or maximum value according as H is positive or negative.

Case 2. H can change sign, since, for small values of α the difference $f(a + h, b + k) - f(a, b)$ and H has the same sign. Therefore, $f(a, b)$ will not be an extreme value.

Case 3. *If* H keeps a constant sign and vanishes for certain values of (h, k), the sign of $f(a+h, b+k) - f(a, b)$ will depend on α, which is an unknown sign and no conclusion can be drawn. This is a doubtful case and further investigation is required.

Let us take first $r \neq 0$. Then H may be written in the following form:

$$H = \frac{(rh + sk)^2 + k^2(rt - s^2)}{r}.$$

1. If $rt - s^2 > 0$, then the numerator of H is the sum of two positive numbers and it never vanishes except when $k = 0$ and $h =$, simultaneously, which is not possible. Hence, H never vanishes and has the same sign as that of r. Thus, $f(a, b)$ will be the minimum or maximum value according as r is positive or negative.

2. If $rt - s^2 < 0$, then the numerator of H may be positive or negative according as $(rh + sk)^2 > k^2(s^2 - rt)$ or $(rh + sk)^2 < k^2(s^2 - rt)$, that is, according to the values of (h, k). Hence, H does not keep the same sign for all values of (h, k). Therefore, $f(a, b)$ is not an extreme value.

3. If $rt - s^2 = 0$, then the numerator of H is a perfect square, but it may vanish for values of (h, k) for which $rh + sk = 0$. Thus, H, without changing sign, may vanish for some values of (h, k). This is a doubtful case in which the sign of $f(a + h, b + k) - f(a, b)$ depends on α and further investigation is required.

   If $r = 0$, then $H = 2shk + tk^2 = k(2sh + tk)$.

4. If $r = 0$ and $s \neq 0$, H vanishes sign with k and $(2sh + tk)$, and there is no extreme value.
5. If $r = 0$ and $s = 0$, H does not change sign, but may vanish when $k = 0$ (without $h = 0$). This is, therefore, a doubtful case and further investigation is required.

Working rule. Let $f(x, y)$ be a function of two variables having the second-order partial derivatives. We want to calculate the maxima and minima of this function. For this purpose, we will follow the following steps:

1. First we calculate $\frac{\partial f}{\partial x}$ and $\frac{\partial f}{\partial y}$ and equate them to zero. That is,

$$\frac{\partial f}{\partial x} = 0 \quad \text{and} \quad \frac{\partial f}{\partial y} = 0.$$

On solving these equations, we get a set of ordered pairs which are stationary points of $f(x, y)$. Let (a, b) be a stationary point.

2. Further, we find

$$r = \frac{\partial^2 f}{\partial x^2} t = \frac{\partial^2 f}{\partial y^2}, s = \frac{\partial^2 f}{\partial x \partial y}.$$

We evaluate these at the point (a, b).

3. Now, we calculate $rt - s^2$ at the point (a, b). Then the following three cases arise:

   a. If $rt - s^2 > 0$ and $r > 0$, then $f(x, y)$ has a minima at the point (a, b) and (a, b) is known as the point of this minima.
   b. If $rt - s^2 > 0$ and $r < 0$, then $f(x, y)$ has a maxima at the point (a, b) and (a, b) is known as the point of this maxima.
   c. If $rt - s^2 < 0$, then $f(x, y)$ has neither maxima nor minima at the point (a, b) and (a, b) is known as the saddle point.
   d. If $rt - s^2 = 0$, then it is a doubtful case. Further investigation is required to determine maxima and minima of the function $f(x, y)$.

Example 1. Find maxima and minima of the function $f(x, y) = x^2 + y^2 + 6x + 12$.

Solution.

$$f(x, y) = x^2 + y^2 + 6x + 12.$$

$$\frac{\partial f}{\partial x} = 2x + 6 \quad \text{and} \quad \frac{\partial f}{\partial y} = 2y.$$

Now, equating to zero these partial derivatives, we get

$$\frac{\partial f}{\partial x} = 2x + 6 = 0 \quad \text{and} \quad \frac{\partial f}{\partial y} = 2y = 0.$$

$$\Rightarrow x = -3 \quad \text{and} \quad y = 0.$$

Now, we investigate maxima and minima at the point $(-3, 0)$. For this purpose, we have

$$r = \frac{\partial^2 f}{\partial x^2} = 2, \quad t = \frac{\partial^2 f}{\partial y^2} = 2, \quad s = \frac{\partial^2 f}{\partial x \partial y} = 0.$$

Now, we evaluate these at the point $(-3, 0)$, so that

$$r = 2, \quad t = 2 \quad \text{and} \quad s = 0.$$

Since $rt - s^2 = 4 > 0$, and $r = 2 > 0$, so f attains its minima at the point $(-3, 0)$. The minimum value of f at the point $(-3, 0)$ is as follows:

$$f_{\min}(-3, 0) = -9.$$

Example 2. Find the maxima and minima of the function $f(x, y) = x^3 + y^3 - 6bxy$.

Solution.

$$f(x, y) = x^3 + y^3 - 6bxy.$$

$$\frac{\partial f}{\partial x} = 3x^2 - 6by \quad \text{and} \quad \frac{\partial f}{\partial y} = 3y^2 - 6bx.$$

Now, equating to zero these partial derivatives, we get

$$\frac{\partial f}{\partial x} = 3x^2 - 6by = 0 \quad \text{and} \quad \frac{\partial f}{\partial y} = 3y^2 - 6bx = 0.$$

On solving these equations, we find that

$$x = 2b \quad \text{and} \quad y = 2b.$$

Now, we investigate the maxima and minima at the point $(2b, 2b)$. For this purpose, we have

$$r = \frac{\partial^2 f}{\partial x^2} = 6x, \quad t = \frac{\partial^2 f}{\partial y^2} = 6y, \quad s = \frac{\partial^2 f}{\partial x \partial y} = -6b.$$

Now, we evaluate these at the point $(2b, 2b)$, so that

$$r = 12b, \quad t = 12b, \quad \text{and} \quad s = -6b.$$

Then $rt - s^2 = 108b^2 > 0$.

Now, $r = 12b$ will be greater than zero if $b > 0$, and less than zero if $b < 0$. Therefore, f will a maximum value if $b < 0$, and a minimum value if $b > 0$.

9.7   Maxima and Minima of Functions of Three Variables Using the Lagrange Method of Undetermined Multipliers

Example 1. Find the minimum value of $u = x^2 + y^2 + z^2$ under the constraint

$$ax + by + cz = m.$$

Solution. Let us define a function F such that

$$F(x, y, z) = x^2 + y^2 + z^2 + \mu(ax + by + cz - m).$$

$$\frac{\partial F}{\partial x} = 2x + \mu a,$$

$$\frac{\partial F}{\partial y} = 2y + \mu b,$$

$$\frac{\partial F}{\partial z} = 2z + \mu c.$$

Now, equating to zero these partial derivatives, we get

$$\frac{\partial F}{\partial x} = 2x + \mu a = 0, \tag{9.15}$$

$$\frac{\partial F}{\partial y} = 2y + \mu b = 0, \tag{9.16}$$

$$\frac{\partial F}{\partial z} = 2z + \mu c = 0. \tag{9.17}$$

Multiplying Eqs. (9.15), (9.16), and (9.17) by x, y and z respectively and adding them, we obtain

$$2(x^2 + y^2 + z^2) + \mu(ax + by + cz) = 0,$$

$$\Rightarrow \mu = -\frac{2u}{m}.$$

Using the value of μ in Eqs. (9.15), (9.16), and (9.17), we get

$$x = \frac{ua}{m}, \quad y = \frac{ub}{m}, \quad \text{and} \quad z = \frac{uc}{m}.$$

Therefore, we have

$$u = x^2 + y^2 + z^2 = \left(\frac{ua}{m}\right)^2 + \left(\frac{ub}{m}\right)^2 + \left(\frac{uc}{m}\right)^2.$$

$$\Rightarrow u = \frac{m^2}{(a^2 + b^2 + c^2)}. \tag{9.18}$$

Now, to discuss the maxima and minima, we will reduce the given function in terms of two variables. Then

$$u = x^2 + y^2 + \frac{1}{c^2}(m - ax - by)^2.$$

$$r = \frac{\partial^2 u}{\partial x^2} = \frac{2}{c^2}(a^2 + c^2).$$

$$t = \frac{\partial^2 u}{\partial y^2} = \frac{2}{c^2}(b^2 + c^2).$$

$$s = \frac{\partial^2 u}{\partial x \partial y} = \frac{2ab}{c^2}.$$

Now, $rt - s^2 = \frac{4}{c^2}(a^2 + b^2 + c^2) > 0$ and $r = \frac{2}{c^2}(a^2 + c^2) > 0$. So, the minima is attained and the value given in Eq. (9.18) is the minimum value of u.

Example 2. Find the maxima and minima of $u = x^2 + y^2 + z^2$ under the constraints $lx^2 + my^2 + nz^2 = 1$ and $px + qy + rz = 0$.

Solution. Let us define a function F such that

$$F(x, y, z) = x^2 + y^2 + z^2 + \mu_1(lx^2 + my^2 + nz^2 - 1)$$
$$+ \mu_2(px + qy + rz).$$
$$\frac{\partial F}{\partial x} = 2x + 2\mu_1 lx + \mu_2 p,$$
$$\frac{\partial F}{\partial y} = 2y + 2\mu_1 my + \mu_2 q,$$
$$\frac{\partial F}{\partial z} = 2z + 2\mu_1 nz + \mu_2 r.$$

Now, equating to zero these partial derivatives, we have

$$\frac{\partial F}{\partial x} = 2x + 2\mu_1 lx + \mu_2 p = 0, \tag{9.19}$$

$$\frac{\partial F}{\partial y} = 2y + 2\mu_1 my + \mu_2 q = 0, \tag{9.20}$$

$$\frac{\partial F}{\partial z} = 2z + 2\mu_1 nz + \mu_2 r = 0. \tag{9.21}$$

Multiplying Eqs. (9.19), (9.20), and (9.21) by x, y and z respectively, and adding them, we get

$$2(x^2 + y^2 + z^2) + 2\mu_1(lx^2 + my^2 + nz^2) + \mu_2(px + qy + rz) = 0.$$
$$2u + 2\mu_1.1 + \mu_2.0 = 0.$$
$$\Rightarrow \mu_1 = -u.$$

Using this value of μ_1 in Eq. (9.19), we have

$$2x - 2ulx + \mu_2 p = 0,$$
$$x = \frac{\mu_2 p}{2(ul - 1)}.$$

Similarly, by using the values of μ_1 in Eqs. (9.20) and (9.21), we get

$$y = \frac{\mu_2 q}{2(um - 1)}.$$

$$z = \frac{\mu_2 r}{2(un - 1)}.$$

Using these values in $px + qy + rz = 0$, we get

$$\frac{\mu_2 p^2}{2(ul - 1)} + \frac{\mu_2 q^2}{2(um - 1)} + \frac{\mu_2 r^2}{2(un - 1)} = 0.$$

$$\Rightarrow \frac{p^2}{(ul - 1)} + \frac{q^2}{(um - 1)} + \frac{r^2}{(un - 1)} = 0.$$

This gives the required values of the maxima and minima.

Example 3. Find the maximum value of $u = xyz$ under the constraint

$$x + y + z = m.$$

Solution. Let us define a function F such that

$$F(x, y, z) = xyz + \mu(x + y + z - m).$$
$$\frac{\partial F}{\partial x} = yz + \mu,$$
$$\frac{\partial F}{\partial y} = xz + \mu,$$
$$\frac{\partial F}{\partial z} = xy + \mu.$$

Now, equating to zero these partial derivatives, we have

$$\frac{\partial F}{\partial x} = yz + \mu = 0, \qquad (9.22)$$

$$\frac{\partial F}{\partial y} = xz + \mu = 0, \qquad (9.23)$$

$$\frac{\partial F}{\partial z} = xy + \mu = 0. \qquad (9.24)$$

Multiplying Eqs. (9.22), (9.23), and (9.24) by x, y and z respectively, and adding them, we get

$$3xyz + \mu(x + y + z) = 0,$$

$$\Rightarrow \mu = -\frac{3u}{m}.$$

Using this value of μ in Eq. (9.22), we get

$$yz = \frac{3u}{m} = \frac{3xyz}{m},$$

$$\Rightarrow x = \frac{m}{3},$$

Similarly, $y = \frac{m}{3}$ and $z = \frac{m}{3}$.

Therefore, we have

$$u = xyz = \frac{m^3}{27}. \tag{9.25}$$

Now, to discuss maxima and minima, we will reduce the given function in terms of two variables. Then

$$u = xy(m - x - y).$$

$$r = \frac{\partial^2 u}{\partial x^2} = -2y.$$

$$t = \frac{\partial^2 u}{\partial y^2} = -2x.$$

$$s = \frac{\partial^2 u}{\partial x \partial y} = m - 2x - 2y.$$

At the point $x = \frac{m}{3}, y = \frac{m}{3}$, and $z = \frac{m}{3}$. We have

$$r = \frac{\partial^2 u}{\partial x^2} = -\frac{2m}{3}.$$

$$t = \frac{\partial^2 u}{\partial y^2} = -\frac{2m}{3}.$$

$$s = \frac{\partial^2 u}{\partial x \partial y} = -\frac{m}{3}.$$

Now, $rt - s^2 = \frac{4m^2}{9} - \frac{m^2}{9} = \frac{m^2}{3} > 0$ and $r = -\frac{2m}{3} < 0$. So, the maxima is attained and the value given in Eq. (9.25) is the maximum value of u.

Exercises

1. Discuss the continuity of the function $f(x, y)$ at $(x, y) = (0, 0)$ where

$$f(x, y) = \begin{cases} \dfrac{x^2 y^2}{x^2 + y^2}, & (x, y) \neq (0, 0), \\ 0, & (x, y) = (0, 0). \end{cases}$$

Ans. Continuous.

2. Discuss the continuity of the function $f(x, y)$ at $(x, y) = (0, 0)$ where

$$f(x, y) = \begin{cases} \dfrac{x^3 y^3}{x^2 + y^2}, & (x, y) \neq (0, 0), \\ 0, & (x, y) = (0, 0). \end{cases}$$

Ans. Continuous.

3. Discuss the continuity of the function $f(x, y)$ at $(x, y) = (0, 0)$ where

$$f(x, y) = \begin{cases} \dfrac{x^2 y}{x^3 + y^3}, & (x, y) \neq (0, 0), \\ 0, & (x, y) = (0, 0). \end{cases}$$

Ans. Discontinuous.

4. Discuss the continuity of the function $f(x, y)$ at $(x, y) = (0, 0)$ where

$$f(x, y) = \begin{cases} \dfrac{x y^3}{x^2 + y^6}, & (x, y) \neq (0, 0), \\ 0, & (x, y) = (0, 0). \end{cases}$$

Ans. Discontinuous.

5. Discuss the continuity of the function $f(x, y)$ at $(x, y) = (0, 0)$ where

$$f(x, y) = \begin{cases} \dfrac{x^2 y}{x^3 + y^3}, & (x, y) \neq (0, 0), \\ 0, & (x, y) = (0, 0). \end{cases}$$

Ans. Continuous.

Chapter 10

Riemann Integration

At elementary stage, the integration is introduced as the inverse of the differentiation. If $h'(x) = f(x)$, for all points in the domain of the function f, then h is called the integral of f. It was Riemann (1782–1867) who first gave the arithmetic concepts of integration free from geometric concepts. This chapter starts with the definition of integrals in the standard way in terms of Riemann sums. Upper and lower integrals are also defined and used in the next section to study the existence of the integral. Further, we discuss some important theorems on Riemann integration. We then discuss Riemann integrability of continuous and monotonic functions. We will see that there are functions that are not Riemann integrable. In the end of this chapter, we present the first and the second forms of the mean value theorem and the fundamental theorem of calculus. It will be noted that all functions will always be considered here as bounded functions.

10.1 Basic Definitions

Partition. Let f be a bounded function on the closed interval $[a, b]$. Then the partition of the closed interval $[a, b]$ is the set of finite points $\{x_0, x_1, x_2, \ldots, x_n\}$ with the property that

$$a = x_0 < x_1 < x_2 < \cdots < x_n = b.$$

The closed subintervals $[x_{i-1}, x_i]$, where $i = 1, 2, 3, \ldots, n$ are called segments and $\Delta x_i = x_i - x_{i-1}$ is called the length of the segment.

Example 1. $P = \left\{0, \frac{1}{2}, 1\right\}$ and $\left\{0, \frac{1}{4}, \frac{1}{2}, \frac{3}{4}, 1\right\}$ are partitions of $[0, 1]$.

Refinement. Let f be a bounded function on the closed interval $[a, b]$. Let P_1 and P_2 be two partitions of $[a, b]$. Then P_1 is called a refinement of P_2 if $P_2 \subset P_1$. That is, the set P_1 contains some extra points from P_2.

Example 1. Let $P_1 = \left\{0, \frac{1}{4}, \frac{1}{2}, \frac{3}{4}, 1\right\}$ and $P_2 = \left\{0, \frac{1}{2}, 1\right\}$ be two partitions of $[0, 1]$. Then $P_2 \subset P_1$, and P_1 is called a refinement of P_2.

Upper and lower Riemann sums. Let f be a bounded function on the closed interval $[a, b]$ and let $P = \{x_0, x_1, x_2, \ldots, x_n\}$ be a partition of $[a, b]$. Let

$$k_i = \inf\{f(x), \ x_{i-1} \le f(x) \le x_i\}, \quad i = 1, 2, 3, \ldots, n$$

and

$$K_i = \sup\{f(x), \ x_{i-1} \le f(x) \le x_i\}, \quad i = 1, 2, 3, \ldots, n.$$

The upper Riemann sum of $f(x)$ with respect to the partition P is given as follows:

$$U(P, f) = \sum_{i=1}^{n} K_i \Delta x_i.$$

The lower Riemann sum of $f(x)$ with respect to the partition P is given as follows:

$$L(P, f) = \sum_{i=1}^{n} k_i \Delta x_i.$$

Theorem 1. *The lower Riemann sum cannot exceed the upper Riemann sum.*

Proof. Let f be a bounded function on the closed interval $[a, b]$ and let $P = \{x_0, x_1, x_2, \ldots, x_n\}$ be a partition of $[a, b]$. Let

$$k_i = \inf\{f(x),\ x_{i-1} \leq f(x) \leq x_i\}, \quad i = 1, 2, 3, \ldots, n$$

and

$$K_i = \sup\{f(x),\ x_{i-1} \leq f(x) \leq x_i\}, \quad i = 1, 2, 3, \ldots, n.$$

We know that

$$k_i \leq K_i \quad \text{for } i = 1, 2, 3, \ldots, n.$$

$$\Rightarrow \quad k_i \Delta x_i \leq K_i \Delta x_i \quad \text{for } i = 1, 2, 3, \ldots, n \ (\text{because } \Delta x_i > 0).$$

$$\Rightarrow \quad \sum_{i=1}^{n} k_i \Delta x_i \leq \sum_{i=1}^{n} K_i \Delta x_i.$$

$$\Rightarrow \quad L(P, f) \leq U(P, f).$$

So, for any partition, the lower Riemann sum cannot exceed the upper Riemann sum.

- Since the function f is bounded on $[a, b]$, therefore, it attains its supremum and infimum on $[a, b]$. Let k be the infimum of f on $[a, b]$ and let K be the supremum of f on $[a, b]$. Now, we have

$$\sum_{i=1}^{n} \Delta x_i = (x_1 - x_0) + (x_2 - x_1) + (x_3 - x_2) + \cdots + (x_n - x_{n-1})$$

$$= x_n - x_0 = b - a.$$

We know that

$$k \leq k_i \leq K_i \leq K \quad \text{for } i = 1, 2, 3, \ldots, n.$$

$$\Rightarrow \quad k \Delta x_i \leq k_i \Delta x_i \leq K_i \Delta x_i \leq K \Delta x_i \quad \text{for } i = 1, 2, 3, \ldots, n.$$

$$\Rightarrow \quad \sum_{i=1}^{n} k \Delta x_i \leq \sum_{i=1}^{n} k_i \Delta x_i \leq \sum_{i=1}^{n} K_i \Delta x_i \leq \sum_{i=1}^{n} K \Delta x_i,$$

$$\Rightarrow \quad k \sum_{i=1}^{n} \Delta x_i \leq \sum_{i=1}^{n} k_i \Delta x_i \leq \sum_{i=1}^{n} K_i \Delta x_i \leq K \sum_{i=1}^{n} \Delta x_i,$$

$$\Rightarrow \quad k(b - a) \leq L(P, f) \leq U(P, f) \leq K(b - a),$$

Theorem 2. *Let f be a bounded function on the closed bounded interval $[a, b]$, and let P_1 be a partition on $[a, b]$. Let P_2 be a refinement of P_1. Then*

$$L(P_1, f) \leq L(P_2, f) \quad and \quad U(P_2, f) \leq U(P_1, f).$$

Proof. Let f be a bounded function on the closed bounded interval $[a, b]$ and let P_1 be a partition on $[a, b]$ given as follows:

$$P_1 = \{a = x_0, x_1, x_2, \ldots, x_{r-1}, x_r, x_{r+1}, \ldots, x_n = b\}.$$

Let us define the refinement P_2 of P_1 by joining a point x_r^* so that

$$P_2 = \{a = x_0, x_1, x_2, \ldots, x_{r-1}, x_r^*, x_r, x_{r+1}, \ldots, x_n = b\}.$$

Let

$$k_i = \inf\{f(x), \ x_{i-1} \leq f(x) \leq x_i\}, \quad i = 1, 2, 3, \ldots, r-1, r, r+1, \ldots, n$$

and

$$K_i = \sup\{f(x), \ x_{i-1} \leq f(x) \leq x_i\},$$
$$i = 1, 2, 3, \ldots, r-1, r, r+1, \ldots, n.$$

The lower Riemann sum with respect to the partition P_1 is given by

$$L(P_1, f) = \sum_{i=1}^{n} k_i \Delta x_i. \tag{10.1}$$

The upper Riemann sum with respect to the partition P_1 is given by

$$U(P_1, f) = \sum_{i=1}^{n} K_i \Delta x_i. \tag{10.2}$$

Let k_r' and k_r'' be the infimum of $f(x)$ on the subintervals $[x_{r-1}, x_r^*]$ and $[x_r^*, x_r]$, respectively, and let K_r' and K_r'' be the supremum of $f(x)$ on the subintervals $[x_{r-1}, x_r^*]$ and $[x_r^*, x_r]$, respectively.

The lower Riemann sum with respect to the partition P_2 is given as follows:

$$L(P_2, f) = \sum_{\substack{i=1 \\ i \neq r}}^{n} k_i \Delta x_i + k_r'(x_r^* - x_{r-1}) + k_r''(x_r - x_r^*). \quad (10.3)$$

The upper Riemann sum with respect to the partition P_2 is given as follows:

$$U(P_2, f) = \sum_{\substack{i=1 \\ i \neq r}}^{n} K_i \Delta x_i + K_r'(x_r^* - x_{r-1}) + K_r''(x_r - x_r^*). \quad (10.4)$$

From the partition, it is clear that

$$k_r = \min\{k_r', k_r''\} \quad \text{and} \quad K_r = \max\{K_r', K_r''\}.$$
$$\Rightarrow \quad k_r \leq k_r', \quad k_r \leq k_r'' \quad \text{and} \quad K_r' \leq K_r, \quad K_r'' \leq K_r.$$

Since $k_r \leq k_r'$ and $k_r \leq k_r''$

$$\Rightarrow \quad k_r(x_r^* - x_{r-1}) \leq k_r'(x_r^* - x_{r-1}) \text{ and } k_r(x_r - x_r^*) \leq k_r''(x_r - x_r^*).$$
$$\Rightarrow \quad k_r(x_r^* - x_{r-1}) + k_r(x_r - x_r^*) \leq k_r'(x_r^* - x_{r-1}) + k_r''(x_r - x_r^*).$$
$$\Rightarrow \quad k_r(x_r - x_{r-1}) \leq k_r'(x_r^* - x_{r-1}) + k_r''(x_r - x_r^*), \quad (10.5)$$

we take $K_r' \leq K_r$, $K_r'' \leq K_r$. Similarly, we can show that

$$K_r'(x_r^* - x_{r-1}) + K_r''(x_r - x_r^*) \leq K_r(x_r - x_{r-1}). \quad (10.6)$$

From Eqs. (10.1), (10.3), and (10.5), we conclude that

$$L(P_1, f) \leq L(P_2, f).$$

From Eqs. (10.2), (10.4), and (10.6), we conclude that

$$U(P_2, f) \leq U(P_1, f).$$

Similarly, we can add a finite number of points in the refinement.

Theorem 3. *Let f be a bounded function on the closed bounded interval $[a, b]$. Let P_1 and P_2 be two partitions on $[a, b]$. Then*

$$L(P_1, f) \leq U(P_2, f) \quad \text{and} \quad L(P_2, f) \leq U(P_1, f).$$

Proof. Let P be a common refinement of P_1 and P_2 defined as $P = P_1 \cup P_2$. We use the relation between upper and lower Riemann sums with respect to the partition P so that

$$L(P, f) \leq U(P, f). \tag{10.7}$$

Since P is a refinement of P_1, by using Theorem 2 above, we can write

$$L(P_1, f) \leq L(P, f) \quad \text{and} \tag{10.8}$$

$$U(P, f) \leq U(P_1, f). \tag{10.9}$$

Since P is a refinement of P_2, by using Theorem 2 above, we can write

$$L(P_2, f) \leq L(P, f) \quad \text{and} \tag{10.10}$$

$$U(P, f) \leq U(P_2, f). \tag{10.11}$$

From Eqs. (10.7), (10.8), and (10.11), we can write

$$L(P_1, f) \leq L(P, f) \leq U(P, f) \leq U(P_2, f).$$
$$\Rightarrow \quad L(P_1, f) \leq U(P_2, f).$$

From Eqs. (10.7), (10.9), and (10.10), we can write

$$L(P_2, f) \leq L(P, f) \leq U(P, f) \leq U(P_1, f).$$
$$\Rightarrow \quad L(P_2, f) \leq U(P_1, f).$$

Upper and lower Riemann integrals. Let $\tau[a, b]$ be the collection of all partitions on $[a, b]$. The infimum of the upper Riemann sum taken over all partitions is called the upper Riemann integral and it is given as follows:

$$\int_a^{-b} f(x)dx = \inf \{U(P, f) : P \in \tau[a, b]\}.$$

The supremum of the lower Riemann sum taken over all partitions is called the lower Riemann integral and it is given as follows:

$$\int_{-a}^b f(x)dx = \sup \{L(P, f) : P \in \tau[a, b]\}.$$

10.2 Riemann Integral

A function f is said to be Riemann integrable if the upper Riemann integral is equal to the lower Riemann integral and the common value of the integral is the value of Riemann integral. That is,

$$\int_a^{-b} f(x)dx = \int_{-a}^b f(x)dx = \int_a^b f(x)dx.$$

If a function defined on $[a, b]$ is Riemann integrable, then we write it as $f \in R[a, b]$.

Theorem 1. *The lower Riemann integral cannot exceed the upper Riemann integral.*

Proof. Let f be a bounded function on $[a, b]$ and let $\tau[a, b]$ be the collection of all partitions on $[a, b]$. Since we know that the lower Riemann sum cannot exceed the upper Riemann sum, we have the following implication:

$$\Rightarrow \quad L(P, f) \leq U(P, f), \quad \forall P \in \tau[a, b].$$

$$\Rightarrow \quad \sup \{L(P, f) : P \in \tau[a, b]\} \leq \inf \{U(P, f) : P \in \tau[a, b]\}.$$

$$\Rightarrow \quad \int_{-a}^b f(x)dx \leq \int_a^{-b} f(x)dx.$$

So, the lower Riemann integral cannot exceed the upper Riemann integral.

Corollary. *Let P_1 be the refinement of P and let it contain r points more than P. Let f be a bounded function on $[a, b]$. That is, $|f(x)| \leq m, \ \forall x \in [a, b]$. Then*

$$L(P, f) \leq L(P_1, f) \leq L(P, f) + 2rm\sigma,$$

$$\text{where } \sigma = \max \Delta x_i (1 \leq i \leq n) \quad \text{and}$$

$$U(P, f) \geq U(P_1, f) \geq U(P, f) - 2rm\sigma.$$

Darboux's theorem. *If f is a bounded function on $[a, b]$, then, for each $\varepsilon > 0$, there corresponds a $\delta > 0$ such that*

a. $L(P, f) > \int_{-a}^{b} f(x)dx - \varepsilon$.

b. $U(P, f) < \int_{a}^{-b} f(x)dx + \varepsilon$.

Here $\sigma(P) < \delta, \ \forall\, P \in \tau[a, b]$.

Proof. Let f be a bounded function on $[a, b]$. That is,

$$|f(x)| \leq m, \quad \forall\, x \in [a, b]. \tag{10.12}$$

Since $\int_{-a}^{b} f(x)dx = \sup\{L(P, f) : P \in \tau[a, b]\}$, for some $\varepsilon > 0$, there exists a partition $P_1 = \{a = x_0, x_1, \ldots, x_r = b\}$ such that

$$L(P_1, f) > \int_{-a}^{b} f(x)dx - \frac{\varepsilon}{2}. \tag{10.13}$$

Let δ be a positive number such that $2(r - 1)m\delta = \dfrac{\varepsilon}{2}$. $\quad$ (10.14)

Let P be any partition with $\sigma(P) < \delta$.

Let P_2 be a refinement of P and P_1. Then $P_2 = P \cup P_1$.

As P_2 is a refinement of P having at most $r - 1$ points more than P (P_1 can have $r - 1$ points different from other partitions because the end points $a = x_0$ and $b = x_r$ are the same for all partitions), by the above corollary we have

$$L(P_2, f) \leq L(P, f) + 2(r - 1)m\sigma. \tag{10.15}$$

Since P_2 is a refinement of P_1,

$$L(P_1, f) \leq L(P_2, f). \tag{10.16}$$

From Eqs. (10.15) and (10.16), we have

$$L(P_1, f) \leq L(P, f) + 2(r - 1)m\sigma. \tag{10.17}$$

From Eqs. (10.13) and (10.17), we get

$$\int_{-a}^{b} f(x)dx - \frac{\varepsilon}{2} < L(P, f) + 2(r - 1)m\sigma \tag{10.18}$$

From Eqs. (10.14) and (10.18), we get

$$\int_{-a}^{b} f(x)dx - \frac{\varepsilon}{2} < L(P, f) + \frac{\varepsilon}{2},$$

$$\Rightarrow \quad \int_{-a}^{b} f(x)dx - \varepsilon < L(P, f).$$

Similarly, we can show that $U(P, f) < \int_{a}^{-b} f(x)dx + \varepsilon$.

10.2.1 *Riemann's necessary and sufficient condition for integrability*

Theorem. *A bounded function $f : [a, b] \to \mathbf{R}$ is Riemann integrable if and only if, for each $\varepsilon > 0$, there exists a partition P such that $U(P, f) - L(P, f) < \varepsilon$.*

Proof. Let us assume that the function $f : [a, b] \to \mathbf{R}$ is Riemann integrable. Then, by the integrability condition:

$$\int_{-a}^{b} f(x)dx = \int_{a}^{-b} f(x)dx, \tag{10.19}$$

we get

$$\int_{a}^{-b} f(x)dx = \inf\{U(P, f) : P \in \tau[a, b]\}.$$

Then, by Darboux's theorem for $\varepsilon > 0$, there exists a partition P_1 such that

$$U(P_1, f) < \int_{a}^{-b} f(x)dx + \frac{\varepsilon}{2}. \tag{10.20}$$

Since

$$\int_{-a}^{b} f(x)dx = \sup\{L(P, f) : P \in \tau[a, b]\},$$

by Darboux's theorem for $\varepsilon > 0$, there exists a partition P_2 such that

$$L(P_2, f) > \int_{-a}^{b} f(x)dx - \frac{\varepsilon}{2}. \tag{10.21}$$

Let $P = P_1 \cup P_2$ be a common refinement of P_1 and P_2. Then

$$U(P, f) \le U(P_1, f) \tag{10.22}$$

and

$$L(P_2, f) \le L(P, f). \tag{10.23}$$

From Eqs. (10.20) and (10.22), we have

$$U(P, f) < \int_a^{-b} f(x)dx + \frac{\varepsilon}{2}. \tag{10.24}$$

From Eqs. (10.21) and (10.23), we have

$$L(P, f) > \int_{-a}^{b} f(x)dx - \frac{\varepsilon}{2}.$$

$$\Rightarrow \quad -L(P, f) < -\int_{-a}^{b} f(x)dx + \frac{\varepsilon}{2}. \tag{10.25}$$

Adding Eqs. (10.24) and (10.25), we get

$$U(P, f) - L(P, f) < \int_a^{-b} f(x)dx - \int_{-a}^{b} f(x)dx + \varepsilon. \tag{10.26}$$

Using Eq. (1) in Eq. (10.26), we get

$$U(P, f) - L(P, f) < \varepsilon.$$

Now, we will prove the converse part. Let us assume that, for each $\varepsilon > 0$, there exists a partition P such that

$$U(P, f) - L(P, f) < \varepsilon. \tag{10.27}$$

Now, we will prove that f is Riemann integrable.

By the definition of the upper Riemann integration, we can write

$$\int_a^{-b} f(x)dx \le U(P, f). \tag{10.28}$$

By the definition of the lower Riemann integration, we can write

$$\int_{-a}^{b} f(x)dx \geq L(P, f).$$

$$\Rightarrow \quad -\int_{-a}^{b} f(x)dx \leq -L(P, f). \tag{10.29}$$

Adding Eqs. (10.28) and (10.29), we have

$$\int_{a}^{-b} f(x)dx - \int_{-a}^{b} f(x)dx \leq U(P, f) - L(P, f). \tag{10.30}$$

By Eqs. (10.27) and (10.30), we get

$$\int_{a}^{-b} f(x)dx - \int_{-a}^{b} f(x)dx < \varepsilon.$$

Since it is true for each $\varepsilon > 0$, we have

$$\int_{a}^{-b} f(x)dx \leq \int_{-a}^{b} f(x)dx. \tag{10.31}$$

Since we know that the lower Riemann integral cannot exceed the upper Riemann integral, we see that

$$\int_{-a}^{b} f(x)dx \leq \int_{a}^{-b} f(x)dx, \tag{10.32}$$

From Eqs. (10.31) and (10.32), we get

$$\int_{-a}^{b} f(x)dx = \int_{a}^{-b} f(x)dx.$$

Example 1. Show that the constant function defined on a closed bounded interval is Riemann integrable.

Solution. Let us assume that the function $f : [a, b] \to \mathbf{R}$ is a constant function defined by

$$f(x) = m, \quad \forall\, x \in [a, b].$$

Let $P = \{a = x_0, x_1, x_2, \ldots, x_n = b\}$ be any arbitrary partition of $[a, b]$ and let

$$k_i = \inf\{f(x), \ x_{i-1} \le f(x) \le x_i\}, \quad i = 1, 2, 3, \ldots, n.$$

and

$$K_i = \sup\{f(x), \ x_{i-1} \le f(x) \le x_i\}, \quad i = 1, 2, 3, \ldots, n.$$

Let $\tau[a, b]$ be the collection of all partitions on $[a, b]$.

Since f is a constant function, we have

$$k_i = m \quad \text{for } i = 1, 2, 3, \ldots, n$$

and

$$K_i = m \quad \text{for } i = 1, 2, 3, \ldots, n.$$

The upper Riemann sum of $f(x)$ with respect to the partition P is given by

$$U(P, f) = \sum_{i=1}^{n} K_i \Delta x_i = K_1 \Delta x_1 + K_2 \Delta x_2 + \cdots + K_n \Delta x_n$$

$$= m(\Delta x_1 + \Delta x_2 + \cdots + \Delta x_n)$$

$$= m(b - a).$$

The lower Riemann sum of $f(x)$ with respect to the partition P is given by

$$L(P, f) = \sum_{i=1}^{n} k_i \Delta x_i = k_1 \Delta x_1 + k_2 \Delta x_2 + \cdots + k_n \Delta x_n$$

$$= m(\Delta x_1 + \Delta x_2 + \cdots + \Delta x_n)$$

$$= m(b - a).$$

Now,

$$\int_a^{-b} f(x) dx = \inf\{U(P, f) : P \in \tau[a, b]\}$$

$$\Rightarrow \quad \int_a^{-b} f(x) dx \le m(b - a). \tag{10.33}$$

Also

$$\int_{-a}^{b} f(x)dx = \sup\{L(P,f) : P \in \tau[a,b]\}$$

$$\Rightarrow \int_{-a}^{b} f(x)dx \geq m(b-a). \tag{10.34}$$

Since we know that

$$\int_{-a}^{b} f(x)dx \leq \int_{a}^{-b} f(x)dx, \tag{10.35}$$

from Eqs. (10.33), (10.34), and (10.35), we get

$$m(b-a) \leq \int_{-a}^{b} f(x)dx \leq \int_{a}^{-b} f(x)dx \leq m(b-a),$$

$$\Rightarrow \int_{-a}^{b} f(x)dx = \int_{a}^{-b} f(x)dx = m(b-a).$$

So, the function f is Riemann integrable and

$$\int_{a}^{b} f(x)dx = m(b-a).$$

Example 2. Show that the function $f(x) = x$ defined on $[0,a]$ is Riemann integrable.

Solution. Let $f : [0,a] \to \mathbf{R}$ be a function defined by

$$f(x) = x, \quad \forall x \in [0,a].$$

Let us take a partition of $[0,a]$ by dividing it into n equal parts so that

$$P = \left\{ 0 = x_0, x_1 = \frac{a}{n}, x_2 = \frac{2a}{n}, \ldots, x_{i-1} = \frac{(i-1)a}{n}, \right.$$

$$\left. x_i = \frac{ia}{n}, \ldots, x_n = \frac{na}{n} \right\}$$

is any arbitrary partition of $[0,a]$ and let

$$k_i = \inf\{f(x), x_{i-1} \leq f(x) \leq x_i\}, \quad i = 1,2,3,\ldots,n$$

and

$$K_i = \sup\{f(x),\ x_{i-1} \le f(x) \le x_i\}, \quad i = 1, 2, 3, \ldots, n.$$

Let $\tau[0,\ a]$ be the collection of all partitions on $[0, a]$.
Now, we have

$$k_i = \frac{(i-1)a}{n} \quad \text{for } i = 1, 2, 3, \ldots, n,$$

$$K_i = \frac{ia}{n} \quad \text{for } i = 1, 2, 3, \ldots, n,$$

And

$$\Delta x_i = \frac{a}{n} \quad \text{for } i = 1, 2, 3, \ldots, n.$$

The upper Riemann sum of $f(x)$ with respect to the partition P is given as follows:

$$U(P, f) = \sum_{i=1}^{n} K_i \Delta x_i = \sum_{i=1}^{n} \frac{ia}{n} \cdot \frac{a}{n} = \frac{a^2}{n^2} \sum_{i=1}^{n} i = \frac{a^2}{n^2} \frac{n(n+1)}{2}$$

$$= \frac{a^2}{2}\left(1 + \frac{1}{n}\right).$$

The lower Riemann sum of $f(x)$ with respect to the partition P is given as follows:

$$L(P, f) = \sum_{i=1}^{n} k_i \Delta x_i = \sum_{i=1}^{n} \frac{(i-1)a}{n} \cdot \frac{a}{n} = \frac{a^2}{n^2} \sum_{i=1}^{n} (i-1) = \frac{a^2}{n^2} \frac{(n-1)n}{2}$$

$$= \frac{a^2}{2}\left(1 - \frac{1}{n}\right).$$

Now,

$$\int_0^{-a} f(x)dx = \inf\{U(P, f) : P \in \tau[0,\ a]\}.$$

$$\Rightarrow \quad \int_0^{-a} f(x)dx \le \frac{a^2}{2}\left(1 + \frac{1}{n}\right) \tag{10.36}$$

and

$$\int_{-0}^{a} f(x)dx = \sup \{L(P, f) : P \in \tau[0, a]\}.$$

$$\Rightarrow \quad \int_{-0}^{a} f(x)dx \geq \frac{a^2}{2}\left(1 - \frac{1}{n}\right). \tag{10.37}$$

Since we know that

$$\int_{-0}^{a} f(x)dx \leq \int_{0}^{-a} f(x)dx, \tag{10.38}$$

from Eqs. (10.36), (10.37), and (10.38), we get

$$\frac{a^2}{2}\left(1 - \frac{1}{n}\right) \leq \int_{-0}^{a} f(x)dx \leq \int_{0}^{-a} f(x)dx \leq \frac{a^2}{2}\left(1 + \frac{1}{n}\right). \tag{10.39}$$

Since P is an arbitrary partition, by letting $n \to \infty$ in Eq. (10.39), we get

$$\int_{-0}^{a} f(x)dx = \int_{0}^{-a} f(x)dx = \frac{a^2}{2}.$$

So, the function f is Riemann integrable and

$$\int_{0}^{a} x\, dx = \frac{a^2}{2}.$$

Example 3. Show that the function $f(x)$ defined on $[0, 1]$ as

$$f(x) = \begin{cases} 2, & x \text{ is rational,} \\ -2, & x \text{ is irrational,} \end{cases}$$

is not Riemann integrable on $[0, 1]$.

Solution. Let $P = \{0 = x_0, x_1, x_2, \ldots, x_n = 1\}$ be any arbitrary partition of $[0, 1]$ and let

$$k_i = \inf\{f(x), x_{i-1} \leq f(x) \leq x_i\}, \quad i = 1, 2, 3, \ldots, n.$$

and

$$K_i = \sup\{f(x), x_{i-1} \le f(x) \le x_i\}, \quad i = 1, 2, 3, \ldots, n.$$

Let $\tau[0, 1]$ be the collection of all partitions on $[0, 1]$.

Since rationals and irrationals are dense in $[0, 1]$, each subinterval $[x_{i-1}, x_i]$ will contain rationals and irrationals. Therefore, we have

$$k_i = -2 \quad \text{for } i = 1, 2, 3, \ldots, n$$

and

$$K_i = 2 \quad \text{for } i = 1, 2, 3, \ldots, n.$$

The upper Riemann sum of $f(x)$ with respect to the partition P is given as follows:

$$U(P, f) = \sum_{i=1}^{n} K_i \Delta x_i = K_1 \Delta x_1 + K_2 \Delta x_2 + \cdots + K_n \Delta x_n$$

$$= 2 \cdot (\Delta x_1 + \Delta x_2 + \cdots + \Delta x_n)$$

$$= 2 \cdot (1 - 0) = 2.$$

The lower Riemann sum of $f(x)$ with respect to the partition P is given as follows:

$$L(P, f) = \sum_{i=1}^{n} k_i \Delta x_i = k_1 \Delta x_1 + k_2 \Delta x_2 + \cdots + k_n \Delta x_n$$

$$= -2 \cdot (\Delta x_1 + \Delta x_2 + \cdots + \Delta x_n)$$

$$= -2 \cdot (1 - 0) = -2.$$

Now,

$$\int_0^{-1} f(x)dx = \inf \{U(P, f) : P \in \tau[0, 1]\}.$$

$$\Rightarrow \quad \int_0^{-1} f(x)dx = 2$$

And

$$\int_{-0}^{1} f(x)dx = \sup\{L(P, f) : P \in \tau[0, 1]\}.$$

$$\Rightarrow \quad \int_{-0}^{1} f(x)dx = -2.$$

Since

$$\int_{-0}^{1} f(x)dx \neq \int_{0}^{-1} f(x)dx,$$

so the function f is not Riemann integrable.

10.3   The Algebra of Riemann Integration

Theorem 1. *If f_1 and f_2 are two Riemann integrable functions on a closed interval $[a, b]$, then their sum $f = f_1 + f_2$ is also Riemann integrable and*

$$\int_{a}^{b} f(x)dx = \int_{a}^{b} f_1(x)dx + \int_{a}^{b} f_2(x)dx.$$

Proof.   Let f_1 and f_2 be two Riemann integrable functions on the closed interval $[a, b]$. Then they are bounded on $[a, b]$. Their sum $f = f_1 + f_2$ will also be bounded, because, if $|f_1(x)| \leq M_1$ and $|f_2(x)| \leq M_2$, then

$$|f(x)| = |(f_1 + f_2)(x)| \leq |f_1(x)| + |f_2(x)| \leq M_1 + M_2.$$

Let $P = \{a = x_0, x_1, x_2, \ldots, x_n = b\}$ be any arbitrary partition of $[a, b]$. Since f_1 is a Riemann integrable function, by Darboux's theorem for $\varepsilon > 0$, there exist a $\delta_1 > 0$ and a partition P with $\sigma(P) < \delta_1$ such that

$$U(P, f_1) - L(P, f_1) < \frac{\varepsilon}{2}. \tag{10.40}$$

Since f_2 is a Riemann integrable function, by Darboux's theorem for $\varepsilon > 0$, there exist a $\delta_2 > 0$ and a partition P with $\sigma(P) < \delta_2$

such that

$$U(P, f_2) - L(P, f_2) < \frac{\varepsilon}{2}. \tag{10.41}$$

Let $\delta = \min\{\delta_1, \delta_2\}$. Then, for $\sigma(P) < \delta$, Eqs. (10.40) and (10.41) can be written as follows:

$$U(P, f_1) - L(P, f_1) < \frac{\varepsilon}{2} \tag{10.42}$$

and

$$U(P, f_2) - L(P, f_2) < \frac{\varepsilon}{2}. \tag{10.43}$$

Let

$$k_i = \inf\{f(x),\ x_{i-1} \le f(x) \le x_i\}, \quad i = 1, 2, 3, \ldots, n.$$
$$K_i = \sup\{f(x),\ x_{i-1} \le f(x) \le x_i\}, \quad i = 1, 2, 3, \ldots, n.$$
$$k_i' = \inf\{f_1(x),\ x_{i-1} \le f_1(x) \le x_i\}, \quad i = 1, 2, 3, \ldots, n.$$
$$K_i' = \sup\{f_1(x),\ x_{i-1} \le f_1(x) \le x_i\}, \quad i = 1, 2, 3, \ldots, n.$$
$$k_i'' = \inf\{f_2(x),\ x_{i-1} \le f_2(x) \le x_i\}, \quad i = 1, 2, 3, \ldots, n.$$
$$K_i'' = \sup\{f_2(x),\ x_{i-1} \le f_2(x) \le x_i\}, \quad i = 1, 2, 3, \ldots, n.$$

Since $f = f_1 + f_2$, so $k_i' + k_i''$ and $K_i' + K_i''$ are the lower and the upper bounds of $f_1(x)$ and $f_2(x)$, respectively, on $[x_{i-1}, x_i]$.

Since k_i and K_i are the infimum and the supremum of $f(x)$ on $[x_{i-1}, x_i]$, $k_i \le K_i$.

By the definition of the infimum and supremum, we can write

$$k_i' + k_i'' \le k_i \le K_i \le K_i' + K_i'' \quad \text{for } i = 1, 2, 3, \ldots, n.$$

Multiplying by $\Delta x_i > 0$ on both sides of the above inequality, we get

$$(k_i' + k_i'')\Delta x_i \le k_i \Delta x_i \le K_i \Delta x_i \le (K_i' + K_i'')\Delta x_i,$$
$$\text{for } i = 1, 2, 3, \ldots, n.$$

$$\Rightarrow \sum_{i=1}^{n} (k_i' + k_i'')\Delta x_i \le \sum_{i=1}^{n} k_i \Delta x_i \le \sum_{i=1}^{n} K_i \Delta x_i \le \sum_{i=1}^{n} (K_i' + K_i'')\Delta x_i.$$

$$\Rightarrow L(P, f_1) + L(P, f_2) \le L(P, f) \le U(P, f) \le U(P, f_1) + U(P, f_2).$$

$$\Rightarrow U(P, f) \le U(P, f_1) + U(P, f_2) \quad \text{and}$$

$$-L(P, f) \le -(L(P, f_1) + L(P, f_2)). \tag{10.44}$$

Adding the above two inequalities, we get

$$U(P, f) - L(P, f) \leq U(P, f_1) + U(P, f_2) - L(P, f_1) - L(P, f_2).$$
$$\Rightarrow \quad U(P, f) - L(P, f) \leq (U(P, f_1) - L(P, f_1)) + (U(P, f_2) - L(P, f_2)).$$
$$(10.45)$$

Now, for $\sigma(P) < \delta$, from Eqs. (3), (4), and (6), we get

$$U(P, f) - L(P, f) < \frac{\varepsilon}{2} + \frac{\varepsilon}{2} = \varepsilon$$
$$\Rightarrow \quad U(P, f) - L(P, f) < \varepsilon.$$

Therefore, the sum function $f = f_1 + f_2$ is also Riemann integrable. Now, we will show the remaining part. Since f_1 and f_2 are Riemann integrable, by Darboux's theorem for $\varepsilon > 0$, there exists a $\delta > 0$ for all partitions for which $\sigma(P) < \delta$ such that

$$L(P, f_1) > \int_a^b f_1(x)dx - \frac{\varepsilon}{2} \quad \text{and} \quad L(P, f_2) > \int_a^b f_2(x)dx - \frac{\varepsilon}{2}.$$
$$(10.46)$$

We know that

$$L(P, f) \leq \int_a^b f(x)dx. \tag{10.47}$$

From Eq. (5), we can write

$$L(P, f_1) + L(P, f_2) \leq L(P, f). \tag{10.48}$$

From Eqs. (10.46), (10.47), and (10.48), we get

$$\int_a^b f_1(x)dx - \frac{\varepsilon}{2} + \int_a^b f_2(x)dx - \frac{\varepsilon}{2} < \int_a^b f(x)dx.$$
$$\Rightarrow \quad \int_a^b f_1(x)dx + \int_a^b f_2(x)dx < \int_a^b f(x)dx + \varepsilon. \tag{10.49}$$

Eq. (10.49) is true for each $\varepsilon > 0$. Therefore, we have

$$\int_a^b f_1(x)dx + \int_a^b f_2(x)dx \leq \int_a^b f(x)dx. \tag{10.50}$$

Since f_1, f_2, and f are Riemann integrable, $-f_1$, $-f_2$, and $-f$ are also Riemann integrable. Then, by Eq. (10.50), we get

$$\int_a^b (-f_1)(x)dx + \int_a^b (-f_2)(x)dx \leq \int_a^b (-f)(x)dx.$$

$$\Rightarrow \quad \int_a^b f_1(x)dx + \int_a^b f_2(x)dx \geq \int_a^b f(x)dx. \tag{10.51}$$

From Eqs. (10.50) and (10.51), we get

$$\int_a^b f_1(x)dx + \int_a^b f_2(x)dx = \int_a^b f(x)dx$$

and

$$\int_a^b f(x)dx = \int_a^b f_1(x)dx + \int_a^b f_2(x)dx.$$

Theorem 2. *If f_1 and f_2 are two Riemann integrable functions on a closed interval $[a, b]$, then their product $f = f_1 \cdot f_2$ is also Riemann integrable.*

Proof.   Let f_1 and f_2 be two Riemann integrable functions on the closed interval $[a, b]$. Then they are bounded on $[a, b]$. Their product $f = f_1 \cdot f_2$ will also be bounded, because, if $|f_1(x)| \leq m$ and $|f_2(x)| \leq m$, then

$$|f(x)| = |(f_1 \cdot f_2)(x)| = |f_1(x)| \cdot |f_2(x)| \leq m^2.$$

For simplicity, we have considered the same bounds for $f_1(x)$ and $f_2(x)$.

Let $P = \{a = x_0, x_1, x_2, \ldots, x_n = b\}$ be any arbitrary partition of $[a, b]$. Since f_1 is a Riemann integrable function, by Darboux's theorem for $\varepsilon > 0$, there exist a $\delta_1 > 0$ and a partition P with $\sigma(P) < \delta_1$ such that

$$U(P, f_1) - L(P, f_1) < \frac{\varepsilon}{2m}. \tag{10.52}$$

Since f_2 is a Riemann integrable function, by Darboux's theorem for $\varepsilon > 0$, there exist a $\delta_2 > 0$ and a partition P with $\sigma(P) < \delta_2$

such that

$$U(P, f_2) - L(P, f_2) < \frac{\varepsilon}{2m}. \qquad (10.53)$$

Let $\delta = \min\{\delta_1, \delta_2\}$. Then, for $\sigma(P) < \delta$, Eqs. (10.52) and (10.53) can be written as follows:

$$U(P, f_1) - L(P, f_1) < \frac{\varepsilon}{2m}. \qquad (10.54)$$

and

$$U(P, f_2) - L(P, f_2) < \frac{\varepsilon}{2m}. \qquad (10.55)$$

Let

$$k_i = \inf\{f(x), \ x_{i-1} \le f(x) \le x_i\}, \quad i = 1, 2, 3, \ldots, n.$$
$$K_i = \sup\{f(x), \ x_{i-1} \le f(x) \le x_i\}, \quad i = 1, 2, 3, \ldots, n.$$
$$k_i' = \inf\{f_1(x), \ x_{i-1} \le f_1(x) \le x_i\}, \quad i = 1, 2, 3, \ldots, n.$$
$$K_i' = \sup\{f_1(x), \ x_{i-1} \le f_1(x) \le x_i\}, \quad i = 1, 2, 3, \ldots, n.$$
$$k_i'' = \inf\{f_2(x), \ x_{i-1} \le f_2(x) \le x_i\}, \quad i = 1, 2, 3, \ldots, n.$$
$$K_i'' = \sup\{f_2(x), \ x_{i-1} \le f_2(x) \le x_i\}, \quad i = 1, 2, 3, \ldots, n.$$

Now, we have

$$
\begin{aligned}
(f_1 f_2)(x_i) - (f_1 f_2)(x_{i-1}) &= f_1(x_i) f_2(x_i) - f_1(x_{i-1}) f_2(x_{i-1}), \\
&= f_1(x_i) f_2(x_i) - f_1(x_i) f_2(x_{i-1}) \\
&\quad + f_1(x_i) f_2(x_{i-1}) - f_1(x_{i-1}) f_2(x_{i-1}) \\
&= f_1(x_i)(f_2(x_i) - f_2(x_{i-1})) \\
&\quad + f_2(x_{i-1})(f_1(x_i) - f_1(x_{i-1})).
\end{aligned}
$$
$$
\begin{aligned}
|(f)(x_i) - (f)(x_{i-1})| &\le |f_1(x_i)| |(f_2(x_i) - f_2(x_{i-1}))| \\
&\quad + |f_2(x_{i-1})| |f_1(x_i) - f_1(x_{i-1})| \\
&\le m(K_i'' - k_i'') + m(K_i' - k_i')
\end{aligned}
$$
$$\Rightarrow \quad (K_i - k_i) \le m(K_i'' - k_i'') + m(K_i' - k_i'), \quad \text{for } i = 1, 2, 3, \ldots, n.$$

Multiplying by $\Delta x_i > 0$ on both sides of the above inequalities, we get

$$\sum_{i=1}^{n} (K_i - k_i)\,\Delta x_i \leq m \sum_{i=1}^{n} (K_i' - k_i')\,\Delta x_i + m \sum_{i=1}^{n} (K_i'' - k_i'')\,\Delta x_i.$$

$$\Rightarrow \quad U(P, f) - L(P, f) \leq m(U(P, f_1) - L(P, f_1))$$
$$+ m(U(P, f_2) - L(P, f_2)). \tag{10.56}$$

Now, for $\sigma(P) < \delta$, from Eqs. (10.54), (10.55), and (10.56), we get

$$U(P, f) - L(P, f) < m \cdot \frac{\varepsilon}{2m} + m \cdot \frac{\varepsilon}{2m} = \varepsilon,$$

$$\Rightarrow \quad U(P, f) - L(P, f) < \varepsilon.$$

Therefore, the product function $f = f_1 \cdot f_2$ is also Riemann integrable.

Theorem 3. *If f_1 and f_2 are two Riemann integrable functions on a closed interval $[a, b]$, and there exists a number $\mu > 0$ such that $f_2(x) \geq \mu,\ \forall\, x \in [a, b]$, then $f = \frac{f_1}{f_2}$ is also Riemann integrable.*

Proof. Let f_1 and f_2 be two Riemann integrable functions on the closed interval $[a, b]$. Then they are bounded on $[a, b]$. Also $f_2(x) \geq \mu$, $\forall\, x \in [a, b]$, and $\mu > 0$. Then $f = \frac{f_1}{f_2}$ will also be bounded, because, if $|f_1(x)| \leq m$ and $\mu \leq |f_2(x)| \leq m$, then

$$|f(x)| = \left| \left(\frac{f_1}{f_2} \right)(x) \right| = \left| \frac{f_1(x)}{f_2(x)} \right| \leq \frac{m}{\mu}.$$

Let $P = \{a = x_0, x_1, x_2, \ldots, x_n = b\}$ be any arbitrary partition of $[a, b]$. Since f_1 is a Riemann integrable function, by Darboux's theorem for $\varepsilon > 0$, there exist a $\delta_1 > 0$ and a partition P with $\sigma(P) < \delta_1$ such that

$$U(P, f_1) - L(P, f_1) < \frac{\varepsilon \mu^2}{2m}. \tag{10.57}$$

Since f_2 is a Riemann integrable function, by Darboux's theorem for $\varepsilon > 0$, there exist a $\delta_2 > 0$ and a partition P with $\sigma(P) < \delta_2$

such that

$$U(P, f_2) - L(P, f_2) < \frac{\varepsilon \mu^2}{2m}. \tag{10.58}$$

Let $\delta = \min\{\delta_1, \delta_2\}$. Then, for $\sigma(P) < \delta$, Eqs. (10.57) and (10.58) can be written as follows:

$$U(P, f_1) - L(P, f_1) < \frac{\varepsilon \mu^2}{2m} \tag{10.59}$$

and

$$U(P, f_2) - L(P, f_2) < \frac{\varepsilon \mu^2}{2m}. \tag{10.60}$$

Let

$$k_i = \inf\{f(x),\ x_{i-1} \le f(x) \le x_i\}, \quad i = 1, 2, 3, \ldots, n.$$
$$K_i = \sup\{f(x),\ x_{i-1} \le f(x) \le x_i\}, \quad i = 1, 2, 3, \ldots, n.$$
$$k_i' = \inf\{f_1(x),\ x_{i-1} \le f_1(x) \le x_i\}, \quad i = 1, 2, 3, \ldots, n.$$
$$K_i' = \sup\{f_1(x),\ x_{i-1} \le f_1(x) \le x_i\}, \quad i = 1, 2, 3, \ldots, n.$$
$$k_i'' = \inf\{f_2(x),\ x_{i-1} \le f_2(x) \le x_i\}, \quad i = 1, 2, 3, \ldots, n.$$
$$K_i'' = \sup\{f_2(x),\ x_{i-1} \le f_2(x) \le x_i\}, \quad i = 1, 2, 3, \ldots, n.$$

Now, we see that

$$\left(\frac{f_1}{f_2}\right)(x_i) - \left(\frac{f_1}{f_2}\right)(x_{i-1}) = \frac{f_1(x_i)}{f_2(x_i)} - \frac{f_1(x_{i-1})}{f_2(x_{i-1})}$$

$$= \frac{1}{f_2(x_i)f_2(x_{i-1})}(f_1(x_i)f_2(x_{i-1}) - f_1(x_{i-1})f_2(x_i)),$$

$$= \frac{1}{f_2(x_i)f_2(x_{i-1})}(f_1(x_i)f_2(x_{i-1}) - f_1(x_i)f_2(x_i) + f_1(x_i)f_2(x_i)$$

$$- f_1(x_{i-1})f_2(x_i))$$

$$= \frac{1}{f_2(x_i)f_2(x_{i-1})}\{f_1(x_i)(f_2(x_{i-1}) - f_2(x_i))$$

$$+ f_2(x_i)(f_1(x_i) - f_1(x_{i-1}))\}$$

and

$$|(f)(x_i) - (f)(x_{i-1})|$$

$$\leq \left| \frac{1}{f_2(x_i)f_2(x_{i-1})} \right| \{|f_1(x_i)||(f_2(x_i) - f_2(x_{i-1}))|$$

$$+ |f_2(x_i)||f_1(x_i) - f_1(x_{i-1})|\}$$

$$\leq \frac{m}{\mu^2}(K_i'' - k_i'') + \frac{m}{\mu^2}(K_i' - k_i')$$

$$\Rightarrow \quad (K_i - k_i) \leq \frac{m}{\mu^2}(K_i'' - k_i'') + \frac{m}{\mu^2}(K_i' - k_i') \quad \text{for } i = 1, 2, 3, \ldots, n.$$

Multiplying by $\Delta x_i > 0$ on both sides of the above inequality, we get

$$\sum_{i=1}^{n}(K_i - k_i)\Delta x_i \leq \frac{m}{\mu^2}\sum_{i=1}^{n}(K_i' - k_i')\Delta x_i + \frac{m}{\mu^2}\sum_{i=1}^{n}(K_i'' - k_i'')\Delta x_i$$

$$\Rightarrow \quad U(P, f) - L(P, f) \leq \frac{m}{\mu^2}(U(P, f_1) - L(P, f_1))$$

$$+ \frac{m}{\mu^2}(U(P, f_2) - L(P, f_2)) \tag{10.61}$$

Now, for $\sigma(P) < \delta$, from Eqs. (10.59), (10.60), and (10.61), we get

$$U(P, f) - L(P, f) < \frac{m}{\mu^2} \cdot \frac{\varepsilon\mu^2}{2m} + \frac{m}{\mu^2} \cdot \frac{\varepsilon\mu^2}{2m} = \varepsilon$$

$$\Rightarrow \quad U(P, f) - L(P, f) < \varepsilon.$$

Therefore, the quotient function $f = \frac{f_1}{f_2}$ is also Riemann integrable.

Theorem 4. *If f is a bounded and Riemann integrable function on $[a, b]$, then $|f|$ is also Riemann integrable on $[a, b]$ and*

$$\left| \int_a^b f dx \right| \leq \int_a^b |f| \, dx.$$

The converse of the above assertion is not true.

Proof. Let f be a bounded and Riemann integrable function on $[a, b]$. Then $|f|$ is also bounded because, if $|f(x)| \leq m$, then

$$||f|(x)| = |(f)(x)| \leq m.$$

Let $P = \{a = x_0, x_1, x_2, \ldots, x_n = b\}$ be any arbitrary partition of $[a, b]$. Since f is a Riemann integrable function, by Darboux's theorem for $\varepsilon > 0$, there exist a $\delta > 0$ and a partition P with $\sigma(P) < \delta$ such that

$$U(P, f) - L(P, f) < \varepsilon. \tag{10.62}$$

Let

$$k_i = \inf\{|f|(x), \ x_{i-1} \leq |f|(x) \leq x_i\}, \quad i = 1, 2, 3, \ldots, n.$$
$$K_i = \sup\{|f|(x), \ x_{i-1} \leq |f|(x) \leq x_i\}, \quad i = 1, 2, 3, \ldots, n.$$
$$k_i' = \inf\{f(x), \ x_{i-1} \leq f(x) \leq x_i\}, \quad i = 1, 2, 3, \ldots, n.$$
$$K_i' = \sup\{f(x), \ x_{i-1} \leq f(x) \leq x_i\}, \quad i = 1, 2, 3, \ldots, n.$$

Now, we have

$$||f|(x_i) - |f|(x_{i-1})| = ||f(x_i)| - |f(x_{i-1})|| \leq |f(x_i) - f(x_{i-1})|$$
$$\leq (K_i' - k_i')$$
$$\Rightarrow \quad (K_i - k_i) \leq (K_i' - k_i') \quad \text{for } i = 1, 2, 3, \ldots, n.$$

Multiplying by $\Delta x_i > 0$ on both sides of the above inequality, we get

$$\sum_{i=1}^{n} (K_i - k_i)\,\Delta x_i \leq \sum_{i=1}^{n} \left(K_i' - k_i'\right)\Delta x_i$$
$$\Rightarrow \quad U(P, |f|) - L(P, |f|) \leq (U(P, f) - L(P, f)). \tag{10.63}$$

Now, for $\sigma(P) < \delta$, from Eqs. (10.62) and (10.63), we get

$$U(P, |f|) - L(P, |f|) < \varepsilon.$$

Therefore, $|f|$ is Riemann integrable.

Now,

$$f(x) \leq |f(x)| = |f|(x), \quad \forall\, x \in [a, b]$$
$$\Rightarrow \quad \int_a^b f\,dx \leq \int_a^b |f|\,dx. \tag{10.64}$$

And

$$-f(x) \leq |f(x)| = |f|(x), \quad \forall\, x \in [a, b].$$

$$\Rightarrow \quad \int_a^b (-f)dx \leq \int_a^b |f|dx$$

$$\Rightarrow \quad -\int_a^b f\,dx \leq \int_a^b |f|dx. \tag{10.65}$$

From Eqs. (10.64) and (10.65), we get

$$\left| \int_a^b f\,dx \right| \leq \int_a^b |f|dx.$$

To show that the converse is not true, we will give an example in which $|f|$ is Riemann integrable, but f is not Riemann integrable. Consider the function $f(x)$ defined on $[0, 1]$ as follows:

$$f(x) = \begin{cases} 1, & x \text{ is rational,} \\ -1, & x \text{ is irrational.} \end{cases}$$

Then f is not Riemann integrable on $[0, 1]$, but $|f|$ is Riemann integrable on $[0, 1]$.

Let $P = \{0 = x_0, x_1, x_2, \ldots, x_n = 1\}$ be any arbitrary partition of $[0, 1]$, and let

$$k_i = \inf\{f(x),\ x_{i-1} \leq f(x) \leq x_i\}, \quad i = 1, 2, 3, \ldots, n.$$
$$K_i = \sup\{f(x),\ x_{i-1} \leq f(x) \leq x_i\}, \quad i = 1, 2, 3, \ldots, n.$$

Let $\tau[0, 1]$ be the collection of all partitions on $[0, 1]$.

Since rationals and irrationals are dense in $[0, 1]$, each subinterval $[x_{i-1}, x_i]$ will contain rationals and irrationals. Therefore, we have

$$k_i = -1 \quad \text{for } i = 1, 2, 3, \ldots, n$$

and

$$K_i = 1 \quad \text{for } i = 1, 2, 3, \ldots, n.$$

The upper Riemann sum of $f(x)$ with respect to the partition P is given as follows:

$$U(P, f) = \sum_{i=1}^{n} K_i \Delta x_i = K_1 \Delta x_1 + K_2 \Delta x_2 + \cdots + K_n \Delta x_n$$

$$= 1 \cdot (\Delta x_1 + \Delta x_2 + \cdots + \Delta x_n)$$

$$= 1 \cdot (1 - 0) = 1.$$

The lower Riemann sum of $f(x)$ with respect to the partition P is given as follows:

$$L(P, f) = \sum_{i=1}^{n} k_i \Delta x_i = k_1 \Delta x_1 + k_2 \Delta x_2 + \cdots + k_n \Delta x_n$$

$$= -1 \cdot (\Delta x_1 + \Delta x_2 + \cdots + \Delta x_n)$$

$$= -1 \cdot (1 - 0) = -1.$$

Now,

$$\int_{0}^{-1} f(x)dx = \inf \{U(P, f) : P \in \tau[a, b]\}.$$

$$\Rightarrow \quad \int_{0}^{-1} f(x)dx = 1$$

and

$$\int_{-0}^{1} f(x)dx = \sup \{L(P, f) : P \in \tau[a, b]\}.$$

$$\Rightarrow \quad \int_{-0}^{1} f(x)dx = -1.$$

Since

$$\int_{-0}^{1} f(x)dx \neq \int_{0}^{-1} f(x)dx,$$

the function f is not Riemann integrable.

Now, we have

$$|f|(x) = |f(x)| = 1 \quad \forall\, x \in [0,1].$$

Let

$$m_i = \inf\{|f|\,(x),\ x_{i-1} \le |f|\,(x) \le x_i\}, \quad i = 1, 2, 3, \ldots, n$$

and

$$M_i = \sup\{|f|\,(x),\ x_{i-1} \le |f|\,(x) \le x_i\}, \quad i = 1, 2, 3, \ldots, n.$$

Therefore, we see that

$$m_i = 1 \quad \text{for } i = 1, 2, 3, \ldots, n.$$
$$M_i = 1 \quad \text{for } i = 1, 2, 3, \ldots, n.$$

The upper Riemann sum of $|f|\,(x)$ with respect to the partition P is given as follows:

$$U(P, f) = \sum_{i=1}^{n} M_i \Delta x_i = 1.$$

The lower Riemann sum of $|f|\,(x)$ with respect to the partition P is given as follows:

$$L(P, f) = \sum_{i=1}^{n} m_i \Delta x_i = 1.$$

Now,

$$\int_0^{-1} f(x)dx = \inf\{U(P, f) : P \in \tau[0,\, 1]\}.$$

$$\Rightarrow \quad \int_0^{-1} f(x)dx = 1$$

And

$$\int_{-0}^{1} f(x)dx = \sup\{L(P, f) : P \in \tau[0,\, 1]\}.$$

$$\Rightarrow \quad \int_{-0}^{1} f(x)dx = 1.$$

Since

$$\int_{-0}^{1} f(x)dx = \int_{0}^{-1} f(x)dx,$$

so the function $|f|$ is Riemann integrable.

Theorem 5. *Every continuous function f defined on $[a,b]$ is Riemann integrable.*

Proof. Let f be a continuous function defined on $[a,b]$. Since every continuous function on a closed interval attains its supremum and infimum on it, it is bounded.

Since every continuous function on a closed interval is also uniformly continuous, f is uniformly continuous and, for each $\varepsilon > 0$, there exists a $\delta > 0$ such that

$$|f(x_1) - f(x_2)| < \frac{\varepsilon}{b-a} \quad \text{whenever } |x_1 - x_2| < \delta \text{ and } x_1, x_2 \in [a,b].$$
(10.66)

Let $P = \{a = x_0, x_1, x_2, \ldots, x_n = b\}$ be a partition of $[a,b]$ with $\sigma(P) < \delta$ and let

$$k_i = \inf\{f(x), \; x_{i-1} \le f(x) \le x_i\}, \quad i = 1, 2, 3, \ldots, n$$

and

$$K_i = \sup\{f(x), \; x_{i-1} \le f(x) \le x_i\}, \quad i = 1, 2, 3, \ldots, n.$$

Since $\sigma(P) < \delta$, from Eq. (10.66), we can write

$$(K_i - k_i) < \frac{\varepsilon}{b-a} \quad \text{for } i = 1, 2, 3, \ldots, n. \tag{10.67}$$

Multiplying by $\Delta x_i > 0$ on both sides of Eq. (10.67), we get

$$\sum_{i=1}^{n} (K_i - k_i)\, \Delta x_i < \frac{\varepsilon}{b-a} \sum_{i=1}^{n} \Delta x_i.$$

$$\Rightarrow \quad U(P,f) - L(P,f) < \frac{\varepsilon}{b-a}(b-a) = \varepsilon.$$

Now, for $\sigma(P) < \delta$,

$$U(P, f) - L(P, f) < \varepsilon,$$

Therefore, the function f is Riemann integrable.

Theorem 6. *Every monotonic function f defined on $[a, b]$ is Riemann integrable.*

Proof. Let f be a monotonic function defined on $[a, b]$ and suppose that f is a monotonic increasing function. If $f(a) = f(b)$, then f is a constant function and it is Riemann integrable. Let $f(a) \neq f(b)$.

Let $P = \{a = x_0, x_1, x_2, \ldots, x_n = b\}$ be a partition of $[a, b]$ with $\sigma(P) < \delta$ and let

$$k_i = \inf\{f(x), \; x_{i-1} \le f(x) \le x_i\}, \quad i = 1, 2, 3, \ldots, n$$

and

$$K_i = \sup\{f(x), \; x_{i-1} \le f(x) \le x_i\}, \quad i = 1, 2, 3, \ldots, n.$$

Since f is a monotonic increasing function, we have

$$k_i = f(x_{i-1}) \quad \text{and} \quad K_i = f(x_i) \quad \text{for } i = 1, 2, 3, \ldots, n$$

and

$$U(P, f) - L(P, f) = \sum_{i=1}^{n} K_i \Delta x_i - \sum_{i=1}^{n} k_i \Delta x_i = \sum_{i=1}^{n} (K_i - k_i) \Delta x_i,$$

$$= \sum_{i=1}^{n} (f(x_i) - f(x_{i-1})) \Delta x_i$$

$$< \frac{\varepsilon}{f(b) - f(a)} \sum_{i=1}^{n} (f(x_i) - f(x_{i-1}))$$

$$= \frac{\varepsilon}{f(b) - f(a)} (f(b) - f(a)) = \varepsilon.$$

So, for $\sigma(P) < \delta$, we get

$$U(P, f) - L(P, f) < \varepsilon,$$

Therefore, the function f is Riemann integrable. Similar proof would apply if f is a monotonic decreasing function. So, every monotonic function is Riemann integrable.

Theorem 7. *If f is a bounded and Riemann integrable function on $[a, b]$ and if k and K are the infimum and supremum of f on $[a, b]$, then*

a. $k(b - a) \leq \int_a^b f(x)dx \leq K(b - a)$, *if $b \geq a$.*
b. $k(b - a) \geq \int_a^b f(x)dx \geq K(b - a)$, *if $b \leq a$.*

Proof. If $a = b$, then the result is obvious.

Let $b > a$ and $P = \{x_0, x_1, x_2, \ldots, x_n\}$ be a partition of $[a, b]$. Let f be a bounded function and consider

$$k_i = \inf \{f(x), x_{i-1} \leq f(x) \leq x_i\}, \quad i = 1, 2, 3, \ldots, n$$

and

$$K_i = \sup \{f(x), x_{i-1} \leq f(x) \leq x_i\}, \quad i = 1, 2, 3, \ldots, n.$$

We know that

$$k \leq k_i \leq K_i \leq K \quad \text{for } i = 1, 2, 3, \ldots, n.$$

$$\Rightarrow \quad k\Delta x_i \leq k_i \Delta x_i \leq K_i \Delta x_i \leq K\Delta x_i \quad \text{for } i = 1, 2, 3, \ldots, n.$$

$$\Rightarrow \quad \sum_{i=1}^n k\Delta x_i \leq \sum_{i=1}^n k_i \Delta x_i \leq \sum_{i=1}^n K_i \Delta x_i \leq \sum_{i=1}^n K\Delta x_i.$$

$$\Rightarrow \quad k\sum_{i=1}^n \Delta x_i \leq \sum_{i=1}^n k_i \Delta x_i \leq \sum_{i=1}^n K_i \Delta x_i \leq K\sum_{i=1}^n \Delta x_i,$$

$$\Rightarrow \quad k(b - a) \leq L(P, f) \leq U(P, f) \leq K(b - a). \tag{10.68}$$

Let $\tau[a, b]$ be the collection of all partitions on $[a, b]$. Then

$$\int_a^{-b} f(x)dx = \inf \{U(P, f) : P \in \tau[a, b]\}$$

and

$$\int_{-a}^{b} f(x)dx = \sup\{L(P, f) : P \in \tau[a, b]\}$$

$$\Rightarrow \quad \int_{a}^{-b} f(x)dx \leq U(P, f) \quad \text{and} \quad \int_{-a}^{b} f(x)dx \geq L(P, f).$$
(10.69)

Also

$$\int_{-a}^{b} f(x)dx \leq \int_{a}^{-b} f(x)dx.$$
(10.70)

From Eqs. (10.68), (10.69), and (10.70), we get

$$k(b-a) \leq L(P, f) \leq \int_{-a}^{b} f(x)dx \leq \int_{a}^{-b} f(x)dx \leq U(P, f) \leq K(b-a).$$

$$\Rightarrow \quad k(b-a) \leq \int_{-a}^{b} f(x)dx \leq \int_{a}^{-b} f(x)dx \leq K(b-a).$$
(10.71)

Since f is a Riemann integrable function on $[a, b]$, we have

$$\int_{-a}^{b} f(x)dx = \int_{a}^{-b} f(x)dx = \int_{a}^{b} f(x)dx.$$
(10.72)

From Eqs. (10.71) and (10.72), we get

$$k(b-a) \leq \int_{a}^{b} f(x)dx \leq K(b-a) \quad \text{if } b \geq a.$$

Let $a > b$. Then, proceeding similarly, we get

$$k(a-b) \leq \int_{b}^{a} f(x)dx \leq K(a-b).$$

$$\Rightarrow \quad -k(b-a) \leq -\int_{a}^{b} f(x)dx \leq -K(b-a).$$

$$\Rightarrow \quad k(b-a) \geq \int_{a}^{b} f(x)dx \geq K(b-a), \quad \text{if } b \leq a.$$

Theorem 8. *If* f *is a bounded and Riemann integrable function on* $[a, b]$, *then* Cf *is also Riemann integrable on* $[a, b]$ *for a constant* C *and*

$$\int_a^b Cf \, dx = C \int_a^b f \, dx.$$

Proof. Let f be a bounded and Riemann integrable function on $[a, b]$. Then Cf is also bounded, because, if $|f(x)| \leq m$, then

$$|Cf(x)| = |C| \, |(f)(x)| \leq |C| \, m.$$

Let $P = \{a = x_0, x_1, x_2, \ldots, x_n = b\}$ be any arbitrary partition of $[a, b]$. Since f is a Riemann integrable function, by Darboux's theorem for $\varepsilon > 0$, there exist a $\delta > 0$ and a partition P with $\sigma(P) < \delta$ such that

$$U(P, f) - L(P, f) < \frac{\varepsilon}{|c|}. \tag{10.73}$$

Let

$$k_i = \inf\{(Cf)(x), \ x_{i-1} \leq (Cf)(x) \leq x_i\}, \quad i = 1, 2, 3, \ldots, n.$$
$$K_i = \sup\{(Cf)(x), \ x_{i-1} \leq (Cf)(x) \leq x_i\}, \quad i = 1, 2, 3, \ldots, n.$$
$$k_i' = \inf\{f(x), \ x_{i-1} \leq f(x) \leq x_i\}, \quad i = 1, 2, 3, \ldots, n.$$
$$K_i' = \sup\{f(x), \ x_{i-1} \leq f(x) \leq x_i\}, \quad i = 1, 2, 3, \ldots, n.$$

Now,

$$|(Cf)(x_i) - (Cf)(x_{i-1})| = |C| \, |f(x_i) - f(x_{i-1})|,$$
$$\leq |C|(K_i' - k_i')$$
$$\Rightarrow \quad (K_i - k_i) \leq |C|(K_i' - k_i') \quad \text{for } i = 1, 2, 3, \ldots, n. \tag{10.74}$$

Multiplying by $\Delta x_i > 0$ on both sides of Eq. (10.74), we get

$$\sum_{i=1}^n (K_i - k_i)\Delta x_i \leq |C| \sum_{i=1}^n (K_i' - k_i')\Delta x_i.$$
$$\Rightarrow \quad U(P, Cf) - L(P, Cf) \leq |C|(U(P, f) - L(P, f)) \tag{10.75}$$

Now, for $\sigma(P) < \delta$, from Eqs. (10.73) and (10.74), we get

$$U(P, Cf) - L(P, Cf) < \varepsilon,$$

Therefore, Cf is Riemann integrable for a constant C.

10.4   Some Important Theorems

Theorem 1 (First mean value theorem). *If f is a Riemann integrable function on $[a, b]$, and k and K are the lower and upper bounds of f on $[a, b]$, respectively, then there exists a number μ with $k \leq \mu \leq K$ such that*

$$\int_a^b f(x) dx = \mu(b - a).$$

Moreover, if f is a continuous function on $[a, b]$, then there exists a number $d \in [a, b]$ such that

$$\int_a^b f(x) dx = f(d)(b - a).$$

Proof. If $a = b$, then the result is obvious.

Let $b > a$ and $P = \{x_0, x_1, x_2, \ldots, x_n\}$ be a partition of $[a, b]$. Suppose that f is a bounded function on $[a, b]$ and let

$$k_i = \inf\{f(x), \ x_{i-1} \leq f(x) \leq x_i\}, \quad i = 1, 2, 3, \ldots, n$$

and

$$K_i = \sup\{f(x), \ x_{i-1} \leq f(x) \leq x_i\}, \quad i = 1, 2, 3, \ldots, n.$$

We know that

$$k \leq k_i \leq K_i \leq K, \quad \text{for } i = 1, 2, 3, \ldots, n.$$

$$\Rightarrow \quad k\Delta x_i \leq k_i \Delta x_i \leq K_i \Delta x_i \leq K \Delta x_i, \quad \text{for } i = 1, 2, 3, \ldots, n.$$

$$\Rightarrow \quad \sum_{i=1}^{n} k\Delta x_i \leq \sum_{i=1}^{n} k_i \Delta x_i \leq \sum_{i=1}^{n} K_i \Delta x_i \leq \sum_{i=1}^{n} K \Delta x_i,$$

$$\Rightarrow \quad k\sum_{i=1}^{n} \Delta x_i \leq \sum_{i=1}^{n} k_i \Delta x_i \leq \sum_{i=1}^{n} K_i \Delta x_i \leq K\sum_{i=1}^{n} \Delta x_i,$$

$$\Rightarrow \quad k(b - a) \leq L(P, f) \leq U(P, f) \leq K(b - a). \tag{10.76}$$

Let $\tau[a, b]$ be the collection of all partitions on $[a, b]$. Then

$$\int_a^{-b} f(x)dx = \inf \{U(P, f) : P \in \tau[a, b]\}$$

and

$$\int_{-a}^b f(x)dx = \sup \{L(P, f) : P \in \tau[a, b]\}.$$

$$\Rightarrow \int_a^{-b} f(x)dx \leq U(P, f) \quad \text{and} \quad \int_{-a}^b f(x)dx \geq L(P, f). \tag{10.77}$$

Also

$$\int_{-a}^b f(x)dx \leq \int_a^{-b} f(x)dx. \tag{10.78}$$

From Eqs. (10.76), (10.77), and (10.78), we get

$$k(b-a) \leq L(P, f) \leq \int_{-a}^b f(x)dx \leq \int_a^{-b} f(x)dx \leq U(P, f) \leq K(b-a).$$

$$\Rightarrow \quad k(b-a) \leq \int_{-a}^b f(x)dx \leq \int_a^{-b} f(x)dx \leq K(b-a). \tag{10.79}$$

Since f is a Riemann integrable function on $[a, b]$, we have

$$\int_{-a}^b f(x)dx = \int_a^{-b} f(x)dx = \int_a^b f(x)dx. \tag{10.80}$$

From Eqs. (10.79) and (10.80), we get

$$k(b-a) \leq \int_a^b f(x)dx \leq K(b-a) \quad \text{if } b > a.$$

$$\Rightarrow \quad k \leq \frac{\int_a^b f(x)dx}{(b-a)} \leq K. \tag{10.81}$$

From Eq. (10.81), it is clear that there is a number lying between k and K which will be equal to $\frac{\int_a^b f(x)dx}{(b-a)}$. Let μ be a number such that

$k \leq \mu \leq K$. Then

$$\frac{\int_a^b f(x)dx}{(b-a)} = \mu.$$

$$\Rightarrow \quad \int_a^b f(x)dx = \mu(b-a). \tag{10.82}$$

If $a > b$, then, by a similar process, we obtain

$$\int_b^a f(x)dx = \mu(a-b).$$

$$\Rightarrow \quad -\int_a^b f(x)dx = -\mu(b-a).$$

$$\Rightarrow \quad \int_a^b f(x)dx = \mu(b-a).$$

Further, if f is a continuous function, then, by intermediate value theorem, it takes on every value between k and K. That is, there exists a number $d \in [a, b]$ such that

$$f(d) = \mu. \tag{10.83}$$

From Eqs. (10.82) and (10.83), we get

$$\int_a^b f(x)dx = f(d)(b-a).$$

Theorem 2 (Second mean value theorem). *Let f and g be two Riemann integrable functions on $[a, b]$, with $g(x) \geq 0$ or $g(x) \leq 0$ $\forall\, x \in [a, b]$. Let k and K be the lower and the upper bounds of f on $[a, b]$, respectively. Then there exists a number μ lying between k and K such that*

$$\int_a^b f(x) \cdot g(x)dx = \mu \int_a^b g(x)dx.$$

Moreover, if f is a continuous function on $[a, b]$, then there exists a number $d \in [a, b]$ such that

$$\int_a^b f(x).g(x)dx = f(d) \int_a^b g(x)dx.$$

Proof. Let $g(x) \geq 0 \,\forall\, x \in [a, b]$. If $g(x) = 0$, then the result is obvious. Let $g(x) > 0 \,\forall\, x \in [a, b]$.

Since f is a Riemann integrable function on $[a, b]$ and since k and K are the lower and the upper bounds of f on $[a, b]$, respectively, we find that

$$k \leq f \leq K, \quad \forall\, x \in [a, b]. \tag{10.84}$$

Since $g(x) > 0 \,\forall\, x \in [a, b]$, upon multiplying it on both sides of Eq. (10.84), we get

$$k \cdot g(x) \leq f(x) \cdot g(x) \leq K \cdot g(x).$$

$$\Rightarrow \quad k \int_a^b g(x)dx \leq \int_a^b f(x) \cdot g(x)dx \leq K \int_a^b g(x)dx.$$

$$\Rightarrow \quad k \leq \frac{\int_a^b f(x) \cdot g(x)dx}{\int_a^b g(x)dx} \leq K. \tag{10.85}$$

From Eq. (10.85), it is clear that there is a number between k and K which will be equal to $\frac{\int_a^b f(x) \cdot g(x)dx}{\int_a^b g(x)dx}$. Let μ be a number such that $k \leq \mu \leq K$. Then

$$\frac{\int_a^b f(x) \cdot g(x)dx}{\int_a^b g(x)dx} = \mu,$$

$$\Rightarrow \quad \int_a^b f(x) \cdot g(x)dx = \mu \int_a^b g(x)dx. \tag{10.86}$$

Further, if f is a continuous function, then, by intermediate value theorem, it takes on every value between k and K. That is, there exists a number $d \in [a, b]$ such that

$$f(d) = \mu. \tag{10.87}$$

From Eqs. (10.86) and (10.87), we get

$$\int_a^b f(x).g(x)dx = f(d) \int_a^b g(x)dx.$$

Theorem 3 (Fundamental theorem of calculus). *If f is a continuous function on $[a,b]$ and h is a differentiable function on $[a,b]$ such that $h'(x) = f(x)$, $\forall\, x \in [a,b]$, then*

$$\int_a^b f(y)dy = h(b) - h(a).$$

Proof. Suppose that f is a continuous function on $[a,b]$. Let its integral $H(x)$ over (a,x) be defined as follows:

$$H(x) = \int_a^x f(y)dy, \quad \forall\, x \in [a,b], \quad \text{and} \quad H(a) = 0. \tag{10.88}$$

Since f is a continuous function on $[a,b]$, it is bounded on $[a,b]$. That is,

$$|f(y)| \le K, \quad \forall\, y \in [a,b]. \tag{10.89}$$

Claim: The function $H(x)$ is continuous.

Let $d \in [a,\, b]$ be an arbitrary point. Then

$$|H(x) - H(d)| = \left| \int_a^x f(y)dy - \int_a^d f(y)dy \right|$$

$$= \left| \int_a^x f(y)dy + \int_d^a f(y)dy \right|,$$

$$= \left| \int_d^x f(y)dy \right| \le \int_d^x |f(y)|\, dy.$$

Using Eq. (10.89), we get

$$|H(x) - H(d)| \le K(x - d). \tag{10.90}$$

Taking $\delta = \frac{\varepsilon}{K}$ and using Eq. (10.90), if $|x - d| < \delta$, then

$$|H(x) - H(d)| < \varepsilon.$$

Since d is an arbitrary point, the function H is continuous on $[a,b]$.

Using the Leibniz rule of differentiation under the sign of integration on both sides of Eq. (10.88), we get

$$H'(x) = \frac{d}{dx}\left(\int_a^x f(y)dy\right).$$

$$\Rightarrow \quad H'(x) = f(x), \quad \forall\, x \in [a, b]. \tag{10.91}$$

$$\text{Given } h'(x) = f(x), \quad \forall\, x \in [a, b], \tag{10.92}$$

from Eqs. (10.91) and (10.92), we obtain

$$H'(x) = h'(x), \quad \forall\, x \in [a, b].$$
$$\Rightarrow \quad H'(x) - h'(x) = 0, \quad \forall\, x \in [a, b].$$
$$\Rightarrow \quad H(x) - h(x) = c, \quad \forall\, x \in [a, b].$$

Therefore, we have

$$H(a) - h(a) = c \tag{10.93}$$

and

$$H(b) - h(b) = c. \tag{10.94}$$

Subtracting into Eq. (10.93) from Eq. (10.94), we get

$$H(b) - H(a) = h(b) - h(a), \tag{10.95}$$

Now, from Eq. (10.88), we have

$$H(b) = \int_a^b f(y)dy, \quad H(a) = 0. \tag{10.96}$$

From Eqs. (10.95) and (10.96), we get

$$\int_a^b f(y)dy = h(b) - h(a).$$

Example 1. Using the definition of Riemann integration, evaluate $\int_0^{\frac{\pi}{2}} \sin x\, dx$.

Solution. Let $f : [0, \frac{\pi}{2}] \to \mathbf{R}$ be a function defined by

$$f(x) = \sin x, \quad \forall\, x \in \left[0, \frac{\pi}{2}\right].$$

Let us take a partition P of $[0, \frac{\pi}{2}]$ by dividing it into n equal parts. So

$$P = \left\{ 0 = x_0, x_1 = \frac{\pi}{2n}, x_2 = \frac{2\pi}{2n}, \ldots, x_{i-1} = \frac{(i-1)\pi}{2n}, \right.$$

$$\left. x_i = \frac{i\pi}{2n}, \ldots, x_n = \frac{n\pi}{2n} \right\}$$

is any arbitrary partition of $[0, \frac{\pi}{2}]$. Also let

$$k_i = \inf\{f(x),\ x_{i-1} \le f(x) \le x_i\}, \quad i = 1, 2, 3, \ldots, n$$

and

$$K_i = \sup\{f(x),\ x_{i-1} \le f(x) \le x_i\}, \quad i = 1, 2, 3, \ldots, n.$$

Let $\tau[0, \frac{\pi}{2}]$ be the collection of all partitions on $[0, \frac{\pi}{2}]$. Since $\sin x$ is a monotonic increasing function in $[0, \frac{\pi}{2}]$,

$$k_i = \sin \frac{(i-1)\pi}{2n} \quad \text{for } i = 1, 2, 3, \ldots, n,$$

$$K_i = \sin \frac{i\pi}{2n} \quad \text{for } i = 1, 2, 3, \ldots, n$$

$$\text{and} \quad \Delta x_i = \frac{\pi}{2n} \quad \text{for } i = 1, 2, 3, \ldots, n.$$

The upper Riemann sum of $f(x)$ with respect to the partition P is given as follows:

$$U(P, f) = \sum_{i=1}^{n} K_i \Delta x_i = \sum_{i=1}^{n} \sin \frac{i\pi}{2n} \cdot \frac{\pi}{2n} = \frac{\pi}{2n} \sum_{i=1}^{n} \sin \frac{i\pi}{2n}.$$

Now,

$$\sum_{r=1}^{n} \sin rx = \frac{\sin\left\{\frac{(n+1)}{2}x\right\} \sin\left(\frac{nx}{2}\right)}{\sin\left(\frac{x}{2}\right)}$$

$$\Rightarrow \quad U(P,f) = \frac{\pi}{2n} \frac{\sin\left\{\frac{(n+1)}{2}\frac{\pi}{2n}\right\} \sin\left(\frac{n}{2}\frac{\pi}{2n}\right)}{\sin\left(\frac{\pi}{4n}\right)},$$

$$\int_{0}^{-\frac{\pi}{2}} \sin x \, dx = \lim_{n\to\infty} U(P,f) = \lim_{n\to\infty} \left\{\frac{2\pi}{4n} \frac{\sin\left\{\frac{(n+1)}{2}\frac{\pi}{2n}\right\} \sin\left(\frac{n}{2}\frac{\pi}{2n}\right)}{\sin\left(\frac{\pi}{4n}\right)}\right\},$$

$$= 2 \cdot \sin\frac{\pi}{4} \lim_{n\to\infty} \left\{\frac{\frac{\pi}{4n}}{\sin\left(\frac{\pi}{4n}\right)}\right\} \lim_{n\to\infty} \sin\left\{\left(1 + \frac{1}{n}\right)\frac{\pi}{4}\right\}$$

$$= 2 \cdot \frac{1}{\sqrt{2}} \cdot 1 \cdot \frac{1}{\sqrt{2}} = 1$$

and

$$\int_{0}^{-\frac{\pi}{2}} \sin x \, dx = 1.$$

The lower Riemann sum of $f(x)$ with respect to the partition P is given as follows:

$$L(P,f) = \sum_{i=1}^{n} k_i \Delta x_i = \sum_{i=1}^{n} \sin\frac{(i-1)\pi}{2n} \cdot \frac{\pi}{2n} = \frac{\pi}{2n} \sum_{i=1}^{n} \sin\frac{(i-1)\pi}{2n},$$

so that

$$L(P,f) = \frac{\pi}{2n} \frac{\sin\left\{\frac{n}{2}\frac{\pi}{2n}\right\} \sin\left(\frac{(n-1)}{2}\frac{\pi}{2n}\right)}{\sin\left(\frac{\pi}{4n}\right)}.$$

Therefore, we have

$$\int_{-0}^{\frac{\pi}{2}} \sin x \, dx = \lim_{n \to \infty} L(P, f) = \lim_{n \to \infty} \left\{ \frac{\pi}{2n} \frac{\sin\left\{ \frac{n}{2} \frac{\pi}{2n} \right\} \sin\left(\frac{(n-1)}{2} \frac{\pi}{2n} \right)}{\sin\left(\frac{\pi}{4n} \right)} \right\},$$

$$= 2 \cdot \sin \frac{\pi}{4} \lim_{n \to \infty} \left\{ \frac{\frac{\pi}{4n}}{\sin\left(\frac{\pi}{4n} \right)} \right\} \lim_{n \to \infty} \sin\left\{ \left(1 - \frac{1}{n} \right) \frac{\pi}{4} \right\}$$

$$= 2 \cdot \frac{1}{\sqrt{2}} \cdot 1 \cdot \frac{1}{\sqrt{2}} = 1$$

and

$$\int_{-0}^{\frac{\pi}{2}} \sin x \, dx = 1.$$

Since

$$\int_{-0}^{\frac{\pi}{2}} \sin x \, dx = \int_{0}^{-\frac{\pi}{2}} \sin x \, dx = 1,$$

so the function f is Riemann integrable and

$$\int_{0}^{\frac{\pi}{2}} \sin x \, dx = 1.$$

Example 2. Show that the function $f(x)$ defined on $[0, 2]$ by

$$f(x) = \begin{cases} x^2 + x^3, & x \text{ is rational,} \\ x^3 + x, & x \text{ is irrational,} \end{cases}$$

is not Riemann integrable on $[0, 2]$.

Solution. Since $\left(x^2 + x^3 \right) - \left(x^3 + x \right) = x^2 - x = x(x - 1)$.

$$\text{so } x^2 + x^3 < x^3 + x \quad \text{if } x \in (0, 1)$$

$$\text{and } x^2 + x^3 > x^3 + x \quad \text{if } x \in (1, 2).$$

Let $P = \{0 = x_0, x_1 = 1, x_2 = 2\}$ be any arbitrary partition of $[0, 2]$ and let

$$k_i = \inf\{f(x), \ x_{i-1} \leq f(x) \leq x_i\}, \quad i = 1, 2$$

and

$$K_i = \sup\{f(x), \ x_{i-1} \leq f(x) \leq x_i\}, \quad i = 1, 2.$$

Since rationals and irrationals are dense in $[0, 2]$, each of the subintervals $[x_{i-1}, x_i]$ contains rationals and irrationals. Therefore, we have

$$k_i = \begin{cases} x^2 + x^3, & \text{if } x \in (0, 1) \\ x^3 + x, & \text{if } x \in (1, 2), \end{cases} \quad i = 1, 2$$

and

$$K_i = \begin{cases} x^3 + x, & \text{if } x \in (0, 1) \\ x^2 + x^3, & \text{if } x \in (1, 2), \end{cases} \quad i = 1, 2.$$

The upper Riemann integration of $f(x)$ with respect to the partition P is given as follows:

$$\overline{\int_0^2} f(x)dx = \int_0^1 f(x)dx + \int_1^2 f(x)dx$$

$$= \int_0^1 (x^3 + x)dx + \int_1^2 (x^2 + x^3)dx,$$

$$\overline{\int_0^2} f(x)dx = \frac{41}{6}.$$

The lower Riemann integration of $f(x)$ with respect to the partition P is given as follows:

$$\underline{\int_{-0}^2} f(x)dx = \int_0^1 f(x)dx + \int_1^2 f(x)dx$$

$$= \int_0^1 (x^2 + x^3)dx + \int_1^2 (x^3 + x)dx,$$

$$\underline{\int_{-0}^2} f(x)dx = \frac{35}{6}.$$

Since

$$\underline{\int_{-0}^2} f(x)dx \neq \overline{\int_0^2} f(x)dx,$$

the function f is not Riemann integrable.

Remark A. If f is a bounded function on $[a, b]$ and if it has a finite number of discontinuous points of f on $[a, b]$, then f is Riemann integrable on $[a, b]$.

Remark B. If f is a bounded function on $[a, b]$ and if A is the set of discontinuous points of f on $[a, b]$, then f is Riemann integrable if A has finite number of limit points.

Example 1. Show that the function $f : [0, 3] \to \mathbf{R}$, defined by $f(x) = x[x]$, $\forall x \in [0, 3]$, where $[x]$ is the greatest integer function, is Riemann integrable and evaluate $\int_0^3 f(x)dx$.

Solution. The function $f : [0, 3] \to \mathbf{R}$ is given by

$$f(x) = \begin{cases} 0 & 0 \le x < 1 \\ x & 1 \le x < 2 \\ 2x & 2 \le x < 3. \end{cases}$$

Therefore, f is discontinuous at $x = 1$, 2 and 3. Since the function f has finite number of discontinuities, it is Riemann integrable by Remark A. Further, we have

$$\int_0^3 f(x)dx = \int_0^1 f(x)dx + \int_1^2 f(x)dx + \int_2^3 f(x)dx$$

$$= \int_0^1 0 \cdot dx + \int_1^2 xdx + \int_2^3 2x\, dx = \frac{13}{2}.$$

Example 2. Show that the function f on the closed interval $[0, 1]$, defined by

$$f(x) = 2rx \quad \text{when} \quad \frac{1}{r+1} < x < \frac{1}{r}, \quad r = 1, 2, 3, \ldots,$$

$$f(0) = 0 \quad \text{and} \quad f\left(\frac{1}{r}\right) = 1,$$

is Riemann integrable and evaluate $\int_0^1 f(x)dx$.

Solution. To test the continuity at general point $x = \frac{1}{r+1}$,

$$\text{L. H. L. } f\left(\frac{1}{r+1} - 0\right) = \lim_{h \to 0} f\left(\frac{1}{r+1} - h\right)$$

$$= \lim_{h \to 0} 2(r+1)\left(\frac{1}{r+1} - h\right) = 2$$

and

$$\text{R. H. L. } f\left(\frac{1}{r+1}+0\right) = \lim_{h \to 0} f\left(\frac{1}{r+1}+h\right)$$

$$= \lim_{h \to 0} 2r\left(\frac{1}{r+1}+h\right) = \frac{2r}{r+1}$$

Since $f\left(\frac{1}{r+1}-0\right) \neq f\left(\frac{1}{r+1}+0\right)$, the function f is not continuous at $x = \frac{1}{r+1}$, where $r = 1, 2, 3, \ldots$.

So, the given function is discontinuous at $= \frac{1}{2}, \frac{1}{3}, \frac{1}{4}, \ldots$.

The set of points of discontinuity is $A = \{\frac{1}{2}, \frac{1}{3}, \frac{1}{4}, \ldots\}$. The set of limit points of A is $\{0\}$, which is finite. So, from Remark B, f is Riemann integrable and

$$\int_{\frac{1}{r+1}}^{1} f(x)dx = \int_{\frac{1}{r+1}}^{\frac{1}{r}} f(x)dx + \int_{\frac{1}{r}}^{\frac{1}{r-1}} f(x)dx + \cdots \int_{\frac{1}{4}}^{\frac{1}{3}} f(x)dx$$

$$+ \int_{\frac{1}{3}}^{\frac{1}{2}} f(x)dx + \int_{\frac{1}{2}}^{1} f(x)dx$$

$$= \int_{\frac{1}{r+1}}^{\frac{1}{r}} 2rx\, dx + \int_{\frac{1}{r}}^{\frac{1}{r-1}} 2(r-1)x\, dx + \cdots \int_{\frac{1}{4}}^{\frac{1}{3}} 6x\, dx$$

$$+ \int_{\frac{1}{3}}^{\frac{1}{2}} 4x\, dx + \int_{\frac{1}{2}}^{1} 2x\, dx$$

$$= 1 + \frac{1}{2^2} + \frac{1}{3^2} + \cdots + \frac{1}{r^2} - \frac{r}{(r+1)^2}.$$

Letting $r \to \infty$ on both sides of the above equation, we get

$$\int_{0}^{1} f(x)dx = \lim_{r \to \infty}\left(1 + \frac{1}{2^2} + \frac{1}{3^2} + \cdots + \frac{1}{r^2} - \frac{r}{(r+1)^2}\right)$$

$$= \sum_{r=1}^{\infty} \frac{1}{r^2} - \lim_{r \to \infty}\left(\frac{r}{(r+1)^2}\right)$$

$$= \sum_{r=1}^{\infty} \frac{1}{r^2} = \frac{\pi^2}{6}.$$

Therefore, we have

$$\int_0^1 f(x)dx = \frac{\pi^2}{6}.$$

Example 3. Show that the function f on $[0, 1]$, defined by

$$f(x) = \frac{1}{2^m} \quad \text{when} \quad \frac{1}{2^{m+1}} < x \leq \frac{1}{2^m}, \quad m = 0, 1, 2, 3, \ldots,$$

$$f(0) = 0,$$

is Riemann integrable and evaluate $\int_0^1 f(x)dx$.

Solution. To test the continuity at a general point $x = \frac{1}{2^m}$, we see that

$$\text{L. H. L. } f\left(\frac{1}{2^m} - 0\right) = \lim_{h \to 0} f\left(\frac{1}{2^m} - h\right) = \lim_{h \to 0}\left(\frac{1}{2^m}\right) = \frac{1}{2^m}$$

and

$$\text{R. H. L. } f\left(\frac{1}{2^m} + 0\right) = \lim_{h \to 0} f\left(\frac{1}{2^m} + h\right) = \lim_{h \to 0}\left(\frac{1}{2^{m-1}}\right) = \frac{1}{2^{m-1}}.$$

Since $f\left(\frac{1}{2^m} - 0\right) \neq f\left(\frac{1}{2^m} + 0\right)$, therefore, the function f is not continuous at $x = \frac{1}{2^m}$, where $m = 0, 1, 2, 3, \ldots$.

So, the given function is discontinuous at $= 1, \frac{1}{2}, \frac{1}{2^2}, \frac{1}{2^3}, \ldots$.

The set of points of discontinuity is $A = \left\{1, \frac{1}{2}, \frac{1}{2^2}, \frac{1}{2^3}, \ldots\right\}$. The set of limit points of A is $\{0\}$, which is finite. So, from Remark B, f is Riemann integrable and

$$\int_{\frac{1}{2^m}}^1 f(x)dx = \int_{\frac{1}{2^m}}^{\frac{1}{2^{m-1}}} f(x)dx + \int_{\frac{1}{2^{m-1}}}^{\frac{1}{2^{m-2}}} f(x)dx + \cdots \int_{\frac{1}{2^3}}^{\frac{1}{2^2}} f(x)dx$$

$$+ \int_{\frac{1}{2^2}}^{\frac{1}{2}} f(x)dx + \int_{\frac{1}{2}}^1 f(x)dx$$

$$= \int_{\frac{1}{2^m}}^{\frac{1}{2^{m-1}}} \frac{1}{2^{m-1}}dx + \int_{\frac{1}{2^{m-1}}}^{\frac{1}{2^{m-2}}} \frac{1}{2^{m-2}}dx + \cdots \int_{\frac{1}{2^3}}^{\frac{1}{2^2}} \frac{1}{2^2}dx$$

$$+ \int_{\frac{1}{2^2}}^{\frac{1}{2}} \frac{1}{2}dx + \int_{\frac{1}{2}}^1 1dx$$

$$= \frac{1}{2^{m-1}} \left(\frac{1}{2^{m-1}} - \frac{1}{2^m} \right) + \frac{1}{2^{m-2}} \left(\frac{1}{2^{m-2}} - \frac{1}{2^{m-1}} \right) \cdots$$

$$+ \frac{1}{2^2} \left(\frac{1}{2^2} - \frac{1}{2^3} \right) + \frac{1}{2} \left(\frac{1}{2} - \frac{1}{2^2} \right) + \left(1 - \frac{1}{2} \right)$$

$$= \frac{1}{2} + \frac{1}{2} \cdot \frac{1}{2^2} + \frac{1}{2^2} \cdot \frac{1}{2^3} + \cdots + \frac{1}{2^{m-1}} \cdot \frac{1}{2^m}$$

$$= \frac{1}{2} \left\{ 1 + \frac{1}{2^2} + \frac{1}{2^4} + \cdots \frac{1}{2^{2m-2}} \right\}$$

$$= \frac{1}{2} \left\{ \frac{1 - \left(\frac{1}{2^2} \right)^m}{1 - \left(\frac{1}{2^2} \right)} \right\} = \frac{2}{3} \left(1 - \frac{1}{4^m} \right).$$

Letting $m \to \infty$ on both sides of the above equation, we get

$$\int_0^1 f(x)dx = \lim_{m \to \infty} \left(\frac{2}{3} \left(1 - \frac{1}{4^m} \right) \right)$$

$$= \frac{2}{3}.$$

Therefore, we have

$$\int_0^1 f(x)dx = \frac{2}{3}.$$

Example 4. Show that the function f on the closed interval $[0, n]$, where n is a natural number, defined by

$$f(x) = 2, \quad \text{when } x \text{ is a natural number,}$$

$$f(0) = 0, \quad \text{otherwise.}$$

is Riemann integrable and evaluate $\int_0^n f(x)dx$.

Solution. The function f is given as follows:

$$f(x) = \begin{cases} 0, & x = 0, 1, 2, \dots, n. \\ 2, & i - 1 < x < i, \quad \text{for } i = 1, 2, \dots, n. \end{cases}$$

The given function is continuous except when x is a natural number. To test the continuity at $x = a$, where a is a natural number, we have

$$\text{L. H. L. } f(a - 0) = \lim_{h \to 0} f(a - h) = \lim_{h \to 0} (2) = 2$$

and

$$\text{R. H. L. } f(a+0) = \lim_{h \to 0} f(a+h) = \lim_{h \to 0} (2) = 2.$$

But $f(a) = 0$. Since $f(a-0) = f(a+0) \neq f(a)$, the function f is not continuous at $x = a$, where $a = 0, 1, 2, \ldots, n$.

So, the given function is discontinuous at $= 0, 1, 2, \ldots, n$.

The set of points of discontinuity is $A = \{0, 1, 2, \ldots, n\}$, which is finite. So, from Remark A, f is Riemann integrable.

$$\int_0^n f(x)dx = \int_0^1 f(x)dx + \int_1^2 f(x)dx + \cdots \int_{n-1}^n f(x)dx,$$

$$= \int_0^1 2 \cdot dx + \int_1^2 2 \cdot dx + \cdots \int_{n-1}^n 2 \cdot dx,$$

$$= \sum_{i=1}^n 2 = 2n.$$

Therefore, we have

$$\int_0^n f(x)dx = 2n.$$

Exercises

1. If $f(x) = x$, $x \in [0, 3]$ and $P = [0, 1, 2, 3]$ is the partition of $[0, 3]$, then evaluate $U(P, f)$ and $L(P, f)$.

   Ans. $U(P, f) = 6$ and $L(P, f) = 3$

2. Show that the function $f(x) = x^3$ defined on $[0, a]$ is Riemann integrable and $\int_0^a x^3\, dx = \frac{a^4}{4}$.

3. Show that the function $f(x)$ defined on $[0, 1]$ by

$$f(x) = \begin{cases} 3, & x \text{ is rational}, \\ -3, & x \text{ is irrational}, \end{cases}$$

   is not Riemann integrable on $[0, 1]$.

4. Using the definition of Riemann integration, evaluate $\int_0^{\frac{\pi}{2}} \cos x\, dx$.

   Ans. 1.

5. Show that the function $f(x)$ defined on $[0,2]$ by

$$f(x) = \begin{cases} 1 + x, & x \text{ is rational,} \\ x^2 + x, & x \text{ is irrational,} \end{cases}$$

is not Riemann integrable on $[0,2]$. Also evaluate $\int_0^{-2} f(x)dx$ and $\int_{-0}^2 f(x)dx$.

Ans. $\frac{16}{3}$ and $\frac{10}{3}$.

6. Show that the function $f(x)$ defined on $[0, \frac{1}{2}]$ by

$$f(x) = \begin{cases} x, & x \text{ is rational,} \\ 1 - x, & x \text{ is irrational,} \end{cases}$$

is not Riemann integrable on $[0, \frac{1}{2}]$. Also evaluate $\int_0^{-\frac{1}{2}} f(x)dx$ and $\int_{-0}^{\frac{1}{2}} f(x)dx$.

Ans. $\frac{3}{8}$ and $\frac{1}{8}$.

7. Show that the function $f : [0,2] \to \mathbf{R}$, defined by $f(x) = x\,[2x]$, $\forall\, x \in [0,2]$, where $[x]$ is the greatest integer function, is Riemann integrable and evaluate $\int_0^2 f(x)dx$.

Ans. $\frac{17}{4}$.

8. Show that the function $f : [0,5] \to \mathbf{R}$, defined by $f(x) = x\,[x]$, $\forall\, x \in [0,5]$, where $[x]$ is the greatest integer function, is Riemann integrable and evaluate $\int_0^5 f(x)dx$.

Ans. 35.

Chapter 11

The Improper Integrals

In the Riemann integral $\int_a^b f(x)dx$ of a function $f(x)$, we assume two conditions: The terminals a and b should be finite and the function $f(x)$ should be bounded in $[a, b]$. In the following examples, the terminal a or b is not finite or the integrand $f(x)$ is not bounded in $[a, b]$:

$$\int_0^\infty x^2 dx, \quad \int_{-\infty}^2 e^{-x} dx, \quad \int_0^2 \frac{1}{x-2} dx.$$

Then we need some new representation of the integral $\int_a^b f(x)dx$, known as an improper integral. It will be assumed throughout the chapter that the number of singular points in any interval is finite.

In this chapter, first we present the type of the improper integrals. In the next section, we discuss some important theorems which include comparison test, μ-test, Abel's test, and Dirichlet's test for the convergence of the improper integral of the first kind. Further, we discuss the absolute convergence of the improper integral of the first kind. In the last section, we discuss some important theorems and tests, which include the p-test, comparison test, μ-test, Abel's test, and Dirichlet's test for the convergence of the improper integrals of the second kind.

11.1   Types of Integrals

On the basis of the above discussion, the integral $\int_a^b f(x)\,dx$ can be divided into the following two parts:

Proper integral. The integral $\int_a^b f(x)dx$ is called a proper integral if a and b are finite and $f(x)$ is bounded in $[a, b]$.

Examples. $\displaystyle\int_0^1 x\,dx, \ \int_1^2 \frac{1}{x^2}dx, \ \int_0^2 e^x dx.$

Remark. Proper integrals always converge.

Improper integral. The integral $\int_a^b f(x)dx$ is called an improper integral if a or b or both are not finite and $f(x)$ is not bounded in $[a, b]$. The improper integral can be divided in three categories:

1. **Improper integral of the first kind.** The integral $\int_a^b f(x)dx$ is called an improper integral of the first kind if a or b or both are not finite, but function $f(x)$ is bounded in $[a, b]$.

Examples. $\displaystyle\int_0^\infty x^2 dx, \ \int_{-\infty}^2 e^x dx.$

2. **Improper integral of the second kind.** The integral $\int_a^b f(x)dx$ is called an improper integral of the second kind if a and b both are finite, but the function $f(x)$ is not bounded in $[a, b]$.

Examples. $\displaystyle\int_0^2 \frac{1}{x-2}dx, \ \int_0^2 \frac{1}{x^2}dx.$

3. **Improper integral of the third kind.** The integral $\int_a^b f(x)dx$ is called an improper integral of the third kind if a or b or both are not finite and also function $f(x)$ is not bounded in $[a, b]$.

Examples. $\displaystyle\int_0^\infty \frac{1}{x^2}dx, \ \int_{-\infty}^2 \frac{1}{x(x-1)}dx, \ \int_{-\infty}^\infty \frac{1}{x-1}dx.$

11.2   Test for the Convergence of an Improper Integral of the First Kind

a. **Test of convergence for the integral $\int_a^\infty f(x)dx$**
If the function $f(x)$ is bounded and integrable for $\forall x \geq a$, then the integral $\int_a^\infty f(x)dx$ is said to be convergent if the limit

$$\lim_{y \to \infty} \int_a^y f(x)dx = \lim_{\varepsilon \to 0} \int_a^{\frac{1}{\varepsilon}} f(x)dx \qquad (11.1)$$

exists. The integral $\int_a^\infty f(x)dx$ is said to be divergent if the limit in Eq. (11.1) is infinity or not unique.

b. **Test of convergence for the integral $\int_{-\infty}^b f(x)dx$**
If the function $f(x)$ is bounded and integrable for $\forall x \leq b$, then the integral $\int_{-\infty}^b f(x)dx$ is said to be convergent if the limit

$$\lim_{y \to -\infty} \int_y^b f(x)dx = \lim_{\delta \to 0} \int_{-\frac{1}{\delta}}^b f(x)dx, \qquad (11.2)$$

exists. The integral $\int_{-\infty}^b f(x)dx$ is said to be divergent if the limit in Eq. (11.2) is infinity or not unique.

c. **Test of convergence for the integral $\int_{-\infty}^\infty f(x)dx$**
If the function $f(x)$ is bounded and integrable for $\forall x$, then the integral $\int_{-\infty}^\infty f(x)dx$ is said to be convergent if the limit

$$\lim_{y \to \infty} \int_a^y f(x)dx + \lim_{y \to -\infty} \int_y^b f(x)dx$$

$$= \lim_{\varepsilon \to 0} \int_a^{\frac{1}{\varepsilon}} f(x)dx + \lim_{\delta \to 0} \int_{-\frac{1}{\delta}}^b f(x)dx \qquad (11.3)$$

exists. The integral $\int_{-\infty}^\infty f(x)dx$ is said to be divergent if the limit in Eq. (11.3) is infinity or not unique.

Example 1. Test the convergence of the integral $\int_0^\infty e^{px} dx$.

Solution.

$$\int_0^\infty e^{px} dx = \lim_{x \to \infty} \int_0^x e^{px} dx$$

$$= \lim_{x \to \infty} \left[\frac{e^{px}}{p} \right]_0^x$$

$$= \frac{1}{p} \lim_{x \to \infty} [e^{px} - 1]$$

$$= \frac{1}{p} [e^\infty - 1] = \infty.$$

Therefore, the integral $\int_0^\infty e^{px} dx$ is divergent.

Example 2. Test the convergence of the integral $\int_{-\infty}^0 e^{-3x} dx$.

Solution.

$$\int_{-\infty}^0 e^{-3x} dx = \lim_{x \to -\infty} \int_x^0 e^{-3x} dx$$

$$= \lim_{x \to -\infty} \left[\frac{e^{-3x}}{-3} \right]_x^0$$

$$= \frac{1}{-3} \lim_{x \to -\infty} [1 - e^{-3x}]$$

$$= \frac{1}{-3} [1 - \infty] = \infty.$$

Therefore, the integral $\int_{-\infty}^0 e^{-3x} dx$ is divergent.

Example 3. Test the convergence of the integral $\int_1^\infty \frac{x}{1+x^2} dx$.

Solution.

$$\int_1^\infty \frac{x}{1+x^2} dx = \lim_{x \to \infty} \int_1^x \frac{x}{1+x^2} dx$$

$$= \frac{1}{2} \lim_{x \to \infty} [\log(1+x^2)]_1^x$$

$$= \frac{1}{2} \lim_{x \to \infty} [\log(1 + x^2) - \log(2)]$$

$$= \frac{1}{2} [\infty - \log(2)] = \infty.$$

Therefore, the integral $\int_1^\infty \frac{x}{1+x^2} dx$ is divergent.

Example 4. Test the convergence of the integral $\int_3^\infty \frac{dx}{(x-2)^2}$.

Solution.

$$\int_3^\infty \frac{dx}{(x-2)^2} = \lim_{x \to \infty} \int_3^x \frac{dx}{(x-2)^2}$$

$$= \lim_{x \to \infty} \left[\frac{-1}{(x-2)} \right]_3^x$$

$$= -\lim_{x \to \infty} \left[\frac{1}{(x-2)} - 1 \right]$$

$$= -[0 - 1] = 1.$$

Therefore, the integral $\int_3^\infty \frac{dx}{(x-2)^2}$ is convergent.

Example 5. Test the convergence of the integral $\int_a^\infty \frac{dx}{x^m}$ when $a > 0$.

Solution.

$$\int_a^\infty \frac{dx}{x^m} = \lim_{x \to \infty} \int_a^x \frac{dx}{x^m}$$

$$= \lim_{\delta \to 0} \int_a^{\frac{1}{\delta}} \frac{dx}{x^m}$$

$$= \lim_{\delta \to 0} \left[\frac{x^{1-m}}{1-m} \right]_a^{\frac{1}{\delta}}$$

$$= \frac{1}{1-m} \lim_{\delta \to 0} \left[\left(\frac{1}{\delta} \right)^{1-m} - a^{1-m} \right]$$

$$= \frac{1}{1-m} \lim_{\delta \to 0} [(\delta)^{m-1} - a^{1-m}]$$

$$= \frac{1}{1-m} \lim_{\delta \to 0} (\delta)^{m-1} - \frac{a^{1-m}}{1-m}.$$

Now, the following three cases arise.

a. When $m - 1 < 0$, that is, $m < 1$, then

$$\lim_{\delta \to 0} (\delta)^{m-1} = \infty$$

and

$$\int_a^\infty \frac{dx}{x^m} = \infty - \frac{a^{1-m}}{1-m} = \infty$$

Therefore, the integral $\int_a^\infty \frac{dx}{x^m}$ is divergent when $m < 1$.
b. When $m - 1 > 0$, that is, $m > 1$, then

$$\lim_{\delta \to 0} (\delta)^{m-1} = 0$$

and

$$\int_a^\infty \frac{dx}{x^m} = 0 - \frac{a^{1-m}}{1-m} = -\frac{a^{1-m}}{1-m}.$$

Therefore, the integral $\int_a^\infty \frac{dx}{x^m}$ is convergent when $m > 1$.
c. When $m = 1$, then

$$\int_a^\infty \frac{dx}{x} = \lim_{x \to \infty} \int_a^x \frac{dx}{x}$$

$$= \lim_{x \to \infty} [\log x]_a^x$$

$$= \lim_{x \to \infty} \log x - \log a = \infty.$$

Therefore, the integral $\int_a^\infty \frac{dx}{x^m}$ is divergent when $m = 1$.

The overall conclusion is that the integral $\int_a^\infty \frac{dx}{x^m}$ is convergent when $m > 1$ and divergent when $m \leq 1$.

Comparison test. *Let $f(x)$ and $h(x)$ be two bounded, positive, and integrable functions on any finite closed interval $[a, b]$. Then*

a. *If $f(x) \leq h(x)$, $\forall x \geq a$, and the integral $\int_a^\infty h(x)dx$ converges, then the integral $\int_a^\infty f(x)dx$ will also converge.*
b. *If $f(x) \geq h(x), \forall x \geq a$, and the integral $\int_a^\infty h(x)dx$ diverges, then the integral $\int_a^\infty f(x)dx$ will also diverge.*

Corollary. *If* $f(x)$ *and* $h(x) \geq 0$ *and* $\lim_{x \to \infty} \frac{f(x)}{h(x)} = l$, *then*

a. *If* l *is a non-zero finite number, then* $\int_a^\infty f(x)dx$ *and* $\int_a^\infty h(x)dx$ *converge or diverge together.*
b. *If* $l = 0$, *then if* $\int_a^\infty h(x)dx$ *converges, then* $\int_a^\infty f(x)dx$ *converges.*
c. *If* $l = \infty$, *then if* $\int_a^\infty h(x)dx$ *diverges, then* $\int_a^\infty f(x)dx$ *diverges.*

Example 1. Test the convergence of the integral $\int_1^\infty \frac{x^{\frac{5}{2}}}{1+x^{\frac{7}{2}}} dx$.

Solution. Let $f(x) = \frac{x^{\frac{5}{2}}}{1+x^{\frac{7}{2}}}$ and $h(x) = \frac{1}{x}$. Then

$f(x)$ and $h(x) > 0$ and $\lim_{x \to \infty} \frac{f(x)}{h(x)} = \lim_{x \to \infty} \frac{x^{\frac{5}{2}}}{1+x^{\frac{7}{2}}} \cdot x = \lim_{x \to \infty} \frac{x^{\frac{7}{2}}}{1+x^{\frac{7}{2}}} = 1$.

Since $\int_1^\infty \frac{1}{x}dx$ diverges, therefore, by the comparison test, the integral $\int_1^\infty \frac{x^{\frac{5}{2}}}{1+x^{\frac{7}{2}}} dx$ also diverges.

Example 2. Test the convergence of the integral $\int_a^\infty \frac{x^{2p}}{1+x^{2q}} dx$, where p and q are positive integers.

Solution. Let $f(x) = \frac{x^{2p}}{1+x^{2q}}$ and $h(x) = \frac{1}{x^{2(q-p)}}$. Then

$$f(x) \quad \text{and} \quad h(x) > 0 \quad \text{and} \quad \lim_{x \to \infty} \frac{f(x)}{h(x)}$$

$$= \lim_{x \to \infty} \frac{x^{2p}}{1+x^{2q}} \cdot x^{2(q-p)} = \lim_{x \to \infty} \frac{x^{2q}}{1+x^{2q}} = 1.$$

Since $\int_a^\infty \frac{1}{x^{2(q-p)}}dx$ converges when $2(q - p) > 1$, and diverges when $2(q - p) \leq 1$, therefore, by the comparison test, the integral $\int_a^\infty \frac{x^{2p}}{1+x^{2q}}dx$ converges when $q - p > \frac{1}{2}$ and diverges when $q - p \leq \frac{1}{2}$.

Example 3. Test the convergence of the integral $\int_2^\infty \frac{x}{\sqrt{1+x^3}} dx$.

Solution. Let $f(x) = \frac{x}{\sqrt{1+x^3}}$ and $h(x) = \frac{1}{x^{\frac{1}{2}}}$. Then

$$f(x) \quad \text{and} \quad h(x) > 0 \quad \text{and} \quad \lim_{x\to\infty} \frac{f(x)}{h(x)} = \lim_{x\to\infty} \frac{x}{\sqrt{1+x^3}} \cdot x^{\frac{1}{2}}$$

$$= \lim_{x\to\infty} \frac{x^{\frac{3}{2}}}{\sqrt{1+x^3}} = \lim_{x\to\infty} \frac{x^{\frac{3}{2}}}{x^{\frac{3}{2}}\sqrt{\frac{1}{x^{\frac{3}{2}}}+1}} = 1.$$

Since $\int_1^\infty \frac{1}{x^{\frac{1}{2}}} dx$ diverges (because $m = \frac{1}{2} < 1$), by the comparison test, the integral $\int_2^\infty \frac{x}{\sqrt{1+x^3}} dx$ also diverges.

μ-test. *Let $f(x)$ be a bounded and integrable function in (a,∞), where $a > 0$, and let $\lim_{x\to\infty} x^\mu f(x) = l$. Then*

a. *The integral $\int_a^\infty f(x) dx$ converges if $\mu > 1$, and l is a finite number.*
b. *The integral $\int_a^\infty f(x) dx$ diverges if $\mu \leq 1$, and l is a non-zero number (can be infinite).*

Example 1. Test the convergence of the integral $\int_3^\infty \frac{x^2}{1+x^2+x^4} dx$.

Solution. Let $f(x) = \frac{x^2}{1+x^2+x^4}$ and $\mu = 4 - 2 = 2$ (highest power in denominator–highest power in numerator). $f(x)$ is bounded in $(3,\infty)$ and

$$\lim_{x\to\infty} x^\mu f(x) = \lim_{x\to\infty} x^2 \cdot \frac{x^2}{1+x^2+x^4} = \lim_{x\to\infty} \frac{x^4}{x^4(\frac{1}{x^4}+\frac{1}{x^2}+1)} = 1.$$

Since $\mu = 2 > 1$, by the μ-test the integral $\int_3^\infty \frac{x^2}{1+x^2+x^4} dx$ converges.

Example 2. Test the convergence of the integral $\int_2^\infty \frac{3x^3}{1+x^{\frac{7}{2}}} dx$.

Solution. Let $f(x) = \frac{3x^3}{1+x^{\frac{7}{2}}}$ and $\mu = \frac{7}{2} - 3 = \frac{1}{2}$. Here $f(x)$ is bounded in $(2, \infty)$ and

$$\lim_{x \to \infty} x^\mu f(x) = \lim_{x \to \infty} x^{\frac{1}{2}} \cdot \frac{3x^3}{1+x^{\frac{7}{2}}} = \lim_{x \to \infty} \frac{3x^{\frac{7}{2}}}{x^{\frac{7}{2}}\left(\frac{1}{x^{\frac{7}{2}}}+1\right)} = 3.$$

Since $\mu = \frac{1}{2} < 1$, by the μ-test, the integral $\int_3^\infty \frac{x^2}{1+x^2+x^4} dx$ diverges.

Example 3. Test the convergence of the integral $\int_1^\infty x^{p-1}e^{-x} dx$.

Solution. Let $f(x) = x^{p-1}e^{-x}$. Then $f(x)$ is bounded in $(1, \infty)$ and $\lim_{x \to \infty} x^\mu f(x) = \lim_{x \to \infty} x^\mu \cdot x^{p-1}e^{-x} = \lim_{x \to \infty} x^{\mu+p-1}e^{-x} = \lim_{x \to \infty} \frac{x^{\mu+p-1}}{e^x} = 0$, for all values of x and p, because exponential power always dominates the polynomials.

Taking $\mu > 1$, by the μ-test, the integral $\int_1^\infty x^{p-1}e^{-x} dx$ converges for all values of x and p.

Example 4. Test the convergence of the integral $\int_0^\infty \frac{x}{(1+x)^3} dx$.

Solution. $\int_0^\infty \frac{x}{(1+x)^3} dx = \int_0^a \frac{x}{(1+x)^3} dx + \int_a^\infty \frac{x}{(1+x)^3} dx = I_1 + I_2$.

Let $f(x) = \frac{x}{(1+x)^3}$ be bounded in $(0, \infty)$. In I_1, the range of integration is finite and $f(x)$ is bounded, so it is a proper integral. Therefore, I_1 converges.

For I_2, $\mu = 3 - 1 = 2$. Now,

$$\lim_{x \to \infty} x^\mu f(x) = \lim_{x \to \infty} x^2 \cdot \frac{x}{(1+x)^3} = \lim_{x \to \infty} \frac{x^3}{x^3\left(\frac{1}{x}+1\right)^3} = 1.$$

Since $\mu = 2 > 1$, by the μ-test, the integral $I_2 = \int_a^\infty \frac{x}{(1+x)^3} dx$ converges.

The integral $\int_0^\infty \frac{x}{(1+x)^3} dx$ is the sum of two convergent integrals, so it is convergent.

Abel's test. *Let $h(x)$ be monotonic and bounded in (a, ∞) and suppose that the integral $\int_a^\infty f(x)dx$ converges. Then the integral $\int_a^\infty f(x)h(x)dx$ of their product also converges.*

Example 1. Test the convergence of the integral $\int_a^\infty \frac{e^{-x}\sin x}{x^2}dx$, where $a > 0$.

Solution. Let $f(x) = \frac{\sin x}{x^2}$ and $h(x) = e^{-x}$.

Now, $\lim_{x\to\infty} h(x) = \lim_{x\to\infty} e^{-x} = 0$.

So, $h(x)$ is a bounded and monotonic decreasing function in (a, ∞) and

$$f(x) = \frac{\sin x}{x^2} \le \frac{1}{x^2}.$$

Since $\int_a^\infty \frac{1}{x^2}dx$ converges, by the comparison test, the integral $\int_a^\infty \frac{\sin x}{x^2}dx$ converges.

Now, by Abel's test, the product-function integral $\int_a^\infty \frac{e^{-x}\sin x}{x^2}dx$ converges.

Example 2. Test the convergence of the integral $\int_a^\infty \frac{(1-e^{-x})\cos x}{x^2}dx$, where $a > 0$.

Solution. Let $f(x) = \frac{\cos x}{x^2}$ and $h(x) = 1 - e^{-x}$.

Now, $\lim_{x\to\infty} h(x) = \lim_{x\to\infty}(1 - e^{-x}) = 1$.

So, $h(x)$ is a bounded and monotonic increasing function in (a, ∞) and

$$f(x) = \frac{\cos x}{x^2} \le \frac{1}{x^2}.$$

Since $\int_a^\infty \frac{1}{x^2}dx$ converges, by the comparison test, the integral $\int_a^\infty \frac{\cos x}{x^2}dx$ converges.

Now, by Abel's test, the product-function integral $\int_a^\infty \frac{(1-e^{-x})\cos x}{x^2}dx$ converges.

Dirichlet's test. *Let $h(x)$ be monotonic and bounded in (a, ∞) with $\lim_{x\to\infty} h(x) = 0$ and suppose that the integral $|\int_a^\infty f(x)dx|$ is bounded for all finite values of x. Then the product-function integral $\int_a^\infty f(x)h(x)dx$ also converges.*

Example 1. Test the convergence of the integral $\int_0^\infty \frac{\sin x}{1+x}dx$.

Solution. We can write

$$\int_0^\infty \frac{\sin x}{1+x}dx = \int_0^1 \frac{\sin x}{1+x}dx + \int_1^\infty \frac{\sin x}{1+x}dx.$$

$\int_0^1 \frac{\sin x}{1+x}dx$ is a proper integral, so it converges. Now, we will check the convergence of $\int_1^\infty \frac{\sin x}{1+x}dx$.

Let $f(x) = \sin x$ and $h(x) = \frac{1}{1+x}$.

Now, $\lim_{x\to\infty} h(x) = \lim_{x\to\infty} \frac{1}{1+x} = 0$ and $h(x)$ is a bounded and monotonic decreasing function in $(1, \infty)$ and $\lim_{x\to\infty} h(x) = 0$. Also

$$\left| \int_1^y \sin x \, dx \right| = |[-\cos x]_1^y| = |-\cos y + \cos 1| \le |\cos y| + |\cos 1| \le 2.$$

So, $|\int_1^y \sin x \, dx|$ is bounded above by 2 for all finite values of y.

Therefore, by Dirichlet's test, the product-function integral $\int_1^\infty \frac{\sin x}{1+x} dx$ converges.

The integral $\int_0^\infty \frac{\sin x}{1+x} dx$ is a sum of two convergent integrals, therefore, it converges.

Example 2. Test the convergence of the integral $\int_0^\infty \frac{e^{-bx} \sin x}{x} dx$.

Solution. We can write

$$\int_0^\infty \frac{e^{-bx} \sin x}{x} dx = \int_0^1 \frac{e^{-bx} \sin x}{x} dx + \int_1^\infty \frac{e^{-bx} \sin x}{x} dx.$$

$\int_0^1 \frac{e^{-bx} \sin x}{x} dx$ is a proper integral, so it converges. Now, we will check the convergence of $\int_1^\infty \frac{e^{-bx} \sin x}{x} dx$.

Let $f(x) = \sin x$ and $h(x) = \frac{e^{-bx}}{x}$.

Now, $\lim_{x\to\infty} h(x) = \lim_{x\to\infty} \frac{e^{-bx}}{x} = 0$ and $h(x)$ is a bounded and monotonic decreasing function in $(1, \infty)$ and $\lim_{x\to\infty} h(x) = 0$. Also

$$\left| \int_1^y \sin x \, dx \right| = |[-\cos x]_1^y| = |-\cos y + \cos 1| \le |\cos y| + |\cos 1| \le 2.$$

So, $|\int_1^y \sin x \, dx|$ is bounded by 2 for all finite values of y.

Therefore, by Dirichlet's test, the product-function integral $\int_1^\infty \frac{e^{-bx} \sin x}{x} dx$ converges.

The integral $\int_0^\infty \frac{e^{-bx} \sin x}{x} dx$, is a sum of two convergent integrals, therefore, it converges.

11.3   Absolute Convergence of Improper Integrals of the First Kind

The improper integral $\int_a^\infty f(x) dx$ is said to be absolutely convergent if $\int_a^\infty |f(x)| dx$ is convergent.

Since $|\int_a^\infty f(x)dx| \leq \int_a^\infty |f(x)|dx$, if $\int_a^\infty f(x)dx$ is absolutely convergent, then it will be convergent, but the converse need not be true.

Example 1. Test the absolute convergence of the integral $\int_a^\infty \frac{\cos x}{x^5}dx$.

Solution. For the absolute convergence

$$\int_a^\infty \left|\frac{\cos x}{x^5}\right| dx \leq \int_a^\infty \left|\frac{1}{x^5}\right| dx \quad (\text{since } |\cos x| \leq 1)$$

$$= \int_a^\infty \frac{1}{x^5}dx \quad \left(\text{since } \frac{1}{x^5} \text{ is positive in } (0,\infty)\right).$$

Now, we will check the convergence of integral $\int_a^\infty \frac{1}{x^5}dx$.

$$\int_a^\infty \frac{1}{x^5}dx = \lim_{y\to\infty}\left[\frac{-1}{4x^4}\right]_a^y = \frac{-1}{4}\lim_{y\to\infty}\left[\frac{1}{y^4} - \frac{1}{a^4}\right] = \frac{1}{4a^4},$$

which is a finite number. So the integral $\int_a^\infty \frac{1}{x^5}dx$ converges. Therefore, the integral $\int_a^\infty \frac{\cos x}{x^5}dx$ converges absolutely.

Example 2. Test the absolute convergence of the integral $\int_0^\infty \frac{\sin mx}{x^2+a^2}dx$.

Solution. For the absolute convergence

$$\int_0^\infty \left|\frac{\sin mx}{x^2+a^2}\right| dx \leq \int_0^\infty \left|\frac{1}{x^2+a^2}\right| dx \quad (\text{since } |\sin mx| \leq 1)$$

$$= \int_0^\infty \frac{1}{x^2+a^2}dx \quad \left(\text{since } \frac{1}{x^2+a^2} \text{ is positive in } (0,\infty)\right).$$

Now, we will check the convergence of the integral $\int_0^\infty \frac{1}{x^2+a^2}dx$.

$$\int_0^\infty \frac{1}{x^2+a^2}dx = \int_0^1 \frac{1}{x^2+a^2}dx + \int_1^\infty \frac{1}{x^2+a^2}dx.$$

Since $\int_0^1 \frac{1}{x^2+a^2}dx$ is a proper integral, it is convergent.

Now, we will check the convergence of integral $\int_1^\infty \frac{1}{x^2+a^2}dx.$ Let $\mu = 2 - 0 = 2$. Then

$$\lim_{x\to\infty} x^2 f(x) = \lim_{x\to\infty} \frac{x^2}{x^2 + a^2} = \lim_{x\to\infty} \frac{x^2}{x^2\left(1 + \left(\frac{a}{x}\right)^2\right)} = 1.$$

By the μ-test, the integral $\int_1^\infty \frac{1}{x^2+a^2}dx$ converges.

Since $\int_0^\infty \frac{1}{x^2+a^2}dx$ is a sum of two convergent integrals, it converges.

So, $\int_0^\infty \frac{\sin mx}{x^2+a^2}dx$ converges absolutely.

Example 3. Test the absolute convergence of the integral $\int_0^\infty \frac{x^{m-1}}{1+x}dx$.

Solution. For the absolute convergence, we see that

$$\int_0^\infty \left|\frac{x^{m-1}}{1+x}\right| dx = \int_0^\infty \frac{x^{m-1}}{1+x}dx, \quad \left(\text{since } \frac{x^{m-1}}{1+x} \text{ is positive in } (0,\infty)\right).$$

Now, we will check convergence of $\int_0^\infty \frac{x^{m-1}}{1+x}dx$.

$$\int_0^\infty \frac{x^{m-1}}{1+x}dx = \int_0^1 \frac{x^{m-1}}{1+x}dx + \int_1^\infty \frac{x^{m-1}}{1+x}dx.$$

$\int_0^1 \frac{x^{m-1}}{1+x}dx$ is a proper integral, so it converges.

Further, we will check the convergence of the integral $\int_1^\infty \frac{x^{m-1}}{1+x}dx$. Let $\mu = 1 - (m - 1) = 2 - m$. Then

$$\lim_{x\to\infty} x^\mu f(x) = \lim_{x\to\infty} x^{2-m}\frac{x^{m-1}}{1+x} = \lim_{x\to\infty} \frac{x}{1+x} = \lim_{x\to\infty} \frac{x}{x\left(\frac{1}{x}+1\right)} = 1.$$

By the μ-test, the integral $\int_1^\infty \frac{x^{m-1}}{1+x}dx$ converges, if $2 - m > 1$, that is, if $m < 1$, and diverges if $2 - m \leq 1$, that is, if $m \geq 1$.

Therefore, the integral $\int_0^\infty \frac{x^{m-1}}{1+x}dx$ converges absolutely if $m < 1$.

11.4   Test for the Convergence of Improper Integrals of the Second Kind

a. Test of convergence for the integral $\int_a^b f(x)dx$ when $f(x)$ is unbounded at $x = a$

If the function $f(x)$ is unbounded at the end-point $x = a$, then the integral $\int_a^b f(x)dx$ is said to be convergent if the following limit:

$$\lim_{\varepsilon \to 0} \int_{a+\varepsilon}^b f(x)dx \tag{11.4}$$

exists. The integral $\int_a^b f(x)dx$ is said to be divergent if the limit in Eq. (11.4) is infinity or not unique.

b. Test of convergence for the integral $\int_a^b f(x)dx$ when $f(x)$ is unbounded at $x = b$

If the function $f(x)$ is unbounded at the end-point $x = b$, then the integral $\int_a^b f(x)dx$ is said to be convergent if the following limit:

$$\lim_{\varepsilon \to 0} \int_a^{b-\varepsilon} f(x)dx \tag{11.5}$$

exists. The integral $\int_a^b f(x)dx$ is said to be divergent if the limit in Eq. (11.5) is infinity or not unique.

c. Test of convergence for the integral $\int_a^b f(x)dx$ when $f(x)$ is unbounded at some interior point $x = c$

If the function $f(x)$ is unbounded at some interior point $a < x = c < b$, then the integral $\int_a^b f(x)dx$ is said to be convergent if

$$\lim_{\varepsilon_1 \to 0} \int_a^{c-\varepsilon_1} f(x)dx + \lim_{\varepsilon_2 \to 0} \int_{c+\varepsilon_2}^b f(x)dx \tag{11.6}$$

exists. The integral $\int_a^b f(x)dx$ is said to be divergent if the limit in Eq. (11.6) is infinity or not unique.

Similar extensions can be listed if the function $f(x)$ is unbounded at more than one interior point.

Remark. The limit in Eq. (11.6) cannot be written as follows:

$$\lim_{\varepsilon \to 0} \left[\int_a^{c-\varepsilon} f(x)dx + \int_{c+\varepsilon}^b f(x)dx \right].$$

Example 1. Test the convergence of the integral $\int_1^2 \frac{dx}{(x-1)^2}$.

Solution.

$$\int_1^2 \frac{dx}{(x-1)^2} = \lim_{\varepsilon \to 0} \int_{1+\varepsilon}^2 \frac{dx}{(x-1)^2}$$

$$= \lim_{\varepsilon \to 0} \left[\frac{-1}{(x-1)} \right]_{1+\varepsilon}^2$$

$$= \lim_{\varepsilon \to 0} \left[-1 + \frac{1}{\varepsilon} \right]$$

$$= -1 + \infty = \infty.$$

Therefore, the integral $\int_1^2 \frac{dx}{(x-1)^2}$ is divergent.

Example 2. Test the convergence of the integral $\int_0^1 \frac{dx}{(1-x)^{\frac{3}{2}}}$.

Solution.

$$\int_0^1 \frac{dx}{(1-x)^{\frac{3}{2}}} = \lim_{\varepsilon \to 0} \int_0^{1-\varepsilon} \frac{dx}{(1-x)^{\frac{3}{2}}}.$$

$$= \lim_{\varepsilon \to 0} \left[\frac{-2}{(1-x)^{\frac{1}{2}}} \right]_0^{1-\varepsilon}$$

$$= \lim_{\varepsilon \to 0} \left[\frac{-2}{(\varepsilon)^{\frac{1}{2}}} + 2 \right]$$

$$= -\infty + 2 = -\infty.$$

Therefore, the integral $\int_0^1 \frac{dx}{(1-x)^{\frac{3}{2}}}$ is divergent.

Example 3. Test the convergence of the integral $\int_{-1}^{1} \frac{dx}{x^3}$.

Solution.

$$\int_{-1}^{1} \frac{dx}{x^3} = \lim_{\varepsilon_1 \to 0} \int_{-1}^{0-\varepsilon_1} \frac{dx}{x^3} + \lim_{\varepsilon_2 \to 0} \int_{0+\varepsilon_2}^{1} \frac{dx}{x^3}$$

$$= \lim_{\varepsilon_1 \to 0} \left[\frac{-1}{2x^2} \right]_{-1}^{0-\varepsilon_1} + \lim_{\varepsilon_2 \to 0} \left[\frac{-1}{2x^2} \right]_{0+\varepsilon_2}^{1}$$

$$= \lim_{\varepsilon_1 \to 0} \left[-\frac{1}{2\varepsilon_1^2} + \frac{1}{2} \right] + \lim_{\varepsilon_2 \to 0} \left[-\frac{1}{2} + \frac{1}{2\varepsilon_2^2} \right]$$

$$= \lim_{\varepsilon_1 \to 0} \left[-\frac{1}{2\varepsilon_1^2} \right] + \lim_{\varepsilon_2 \to 0} \left[\frac{1}{2\varepsilon_2^2} \right]$$

Since both limits do not exist, the integral $\int_{-1}^{1} \frac{dx}{x^3}$ is divergent.

Theorem 1 (p-test). *The integral $\int_{a}^{b} \frac{dx}{(x-a)^p}$, converges when $p < 1$, and diverges when $p \geq 1$.*

Proof. Since $f(x) = \frac{1}{(x-a)^p}$ is unbounded at the end-point $x = a$, for convergence, we can write

$$\int_{a}^{b} \frac{dx}{(x-a)^p} = \lim_{\varepsilon \to 0} \int_{a+\varepsilon}^{b} \frac{dx}{(x-a)^p}$$

$$= \lim_{\varepsilon \to 0} \left[\frac{(x-a)^{1-p}}{1-p} \right]_{a+\varepsilon}^{b}$$

$$= \frac{1}{1-p} \lim_{\varepsilon \to 0} [(b-a)^{1-p} - (\varepsilon)^{1-p}]$$

$$= \frac{(b-a)^{1-p}}{1-p} - \frac{1}{1-p} \lim_{\varepsilon \to 0} \varepsilon^{1-p}.$$

Now, the following cases arise:

Case 1. When $1 - p > 0$, that is, $p < 1$, then $\lim_{\varepsilon \to 0} \varepsilon^{1-p} = \infty$.

$$\int_a^b \frac{dx}{(x-a)^p} = \lim_{\varepsilon \to 0} \int_{a+\varepsilon}^b \frac{dx}{(x-a)^p} = \frac{(b-a)^{1-p}}{1-p} - \frac{1}{1-p}.0$$

$$= \frac{(b-a)^{1-p}}{1-p},$$

which is a finite number.

So, the integral $\int_a^b \frac{dx}{(x-a)^p}$ converges when $p < 1$.

Case 2. When $1 - p < 0$, that is, $p > 1$, then $\lim_{\varepsilon \to 0} \varepsilon^{1-p} = \infty$.

$$\int_a^b \frac{dx}{(x-a)^p} = \lim_{\varepsilon \to 0} \int_{a+\varepsilon}^b \frac{dx}{(x-a)^p} = \frac{(b-a)^{1-p}}{1-p} - \frac{1}{1-p}.\infty$$

$$= \infty.$$

So, the integral $\int_a^b \frac{dx}{(x-a)^p}$ diverges when $p > 1$.

Case 3. When $1 - p = 0$, that is, $p = 1$, then

$$\int_a^b \frac{dx}{x-a} = \lim_{\varepsilon \to 0} \int_{a+\varepsilon}^b \frac{dx}{x-a}$$

$$= \lim_{\varepsilon \to 0} [\log(x-a)]_{a+\varepsilon}^b$$

$$= \lim_{\varepsilon \to 0} [\log(b-a) - \log(\varepsilon)]$$

$$= \log(b-a) - \lim_{\varepsilon \to 0} [\log(\varepsilon)]$$

$$= \log(b-a) - (-\infty) = \infty.$$

So, the integral $\int_a^b \frac{dx}{(x-a)^p}$ diverges when $p = 1$.

Therefore, the integral $\int_a^b \frac{dx}{(x-a)^p}$ converges when $p < 1$ and diverges when $p \geq 1$.

Theorem 2 (p-test). $\int_a^b \frac{dx}{(b-x)^p}$, *converges when $p < 1$ and diverges when $p \geq 1$.*

Proof. Similar as in Theorem 1 above.

Comparison test for improper integral of the second kind.
Let $f(x)$ and $h(x)$ be two positive and integrable functions on $(a, b]$ and suppose that both are unbounded at $x = a$. Then

a. *If $f(x) \leq h(x), \forall x \in (a, b]$, and the integral $\int_a^b h(x)dx$ converges, then the integral $\int_a^b f(x)dx$ will also converge.*

b. *If $f(x) \geq h(x), \forall x \in (a, b]$, and the integral $\int_a^b h(x)dx$ diverges, then the integral $\int_a^b f(x)dx$ will also diverge.*

Corollary. *If $f(x)$ and $h(x) \geq 0$, both are unbounded at $x = a$ and $lim_{x \to a} \frac{f(x)}{h(x)} = l$, then*

a. *If l is a non-zero finite number, then the integrals $\int_a^b f(x)dx$ and $\int_a^b h(x)dx$ converge or diverge together.*

b. *If $l = 0$, then if $\int_a^b h(x)dx$ converges, the integral $\int_a^b f(x)dx$ converges.*

c. *If $l = \infty$, then if $\int_a^b h(x)dx$ diverges, the integral $\int_a^b f(x)dx$ diverges.*

Example 1. Test convergence of the integral $\int_0^1 \frac{1}{\sqrt{x(x+1)}}$.

Solution. Let $f(x) = \frac{1}{\sqrt{x(x+1)}}$ and $h(x) = \frac{1}{\sqrt{x}}$.
Then both $f(x)$ and $h(x)$ are unbounded at $x = 0$.

$$\lim_{x \to 0} \frac{f(x)}{h(x)} = \lim_{x \to 0} \frac{1}{\sqrt{x(x+1)}} \cdot \sqrt{x}$$

$$= \lim_{x \to 0} \frac{1}{\sqrt{x}\sqrt{(1+x)}} \cdot \sqrt{x}$$

$$= \lim_{x \to 0} \frac{1}{\sqrt{(1+x)}} = 1.$$

Since $\int_0^1 \frac{1}{\sqrt{x}}$ converges by the p-test because $p = \frac{1}{2} < 1$, the integral $\int_0^1 \frac{1}{\sqrt{x(x+1)}}$ converges by the comparison test.

Example 2. Test convergence of the integral $\int_{\frac{1}{2}}^1 \frac{1}{x^{\frac{5}{2}}(1-x)}$.

Solution. Let $f(x) = \dfrac{1}{x^{\frac{5}{2}}(1-x)}$ and $h(x) = \dfrac{1}{(1-x)}$.

Then both $f(x)$ and $h(x)$ are unbounded at $x = 1$.

$$\lim_{x \to 1} \frac{f(x)}{h(x)} = \lim_{x \to 1} \frac{1}{x^{\frac{5}{2}}(1-x)} \cdot (1-x)$$

$$= \lim_{x \to 1} \frac{1}{x^{\frac{5}{2}}} = 1$$

Since $\int_{\frac{1}{2}}^{1} \frac{1}{(1-x)}$ converges by the p-test because $p = \frac{1}{2} < 1$, the integral $\int_{\frac{1}{2}}^{1} \frac{1}{x^{\frac{5}{2}}(1-x)}$ converges by the comparison test.

Gamma function. The definite integral $\int_0^\infty x^{m-1}e^{-x}dx$, for $m > 0$, is known as the Gamma function and is denoted by $\lceil(m)$. Thus,

$$\lceil(m) = \int_0^\infty x^{m-1}e^{-x}dx, \quad m > 0.$$

Example 3 (Convergence of the gamma-function integral). Show that the integral $\int_0^\infty x^{m-1}e^{-x}dx$ converges if $m > 0$.

Solution. We can write

$$\int_0^\infty x^{m-1}e^{-x}dx = \int_0^1 x^{m-1}e^{-x}dx + \int_1^\infty x^{m-1}e^{-x}dx.$$

This integral is a combination of both the first-kind and second-kind improper integrals. First, we will check the convergence of $\int_0^1 x^{m-1}e^{-x}dx$. Since a given function $f(x) = x^{m-1}e^{-x}$ can be unbounded at $x = 0$, the following cases arise:

Case 1. If $m > 1$, then

$$\lim_{x \to 0} x^{m-1}e^{-x} = 0.$$

So, the function $f(x) = x^{m-1}e^{-x}$ is bounded in $[0, 1]$. Therefore, the integral $\int_0^1 x^{m-1}e^{-x}dx$ is a proper integral and converges for $m > 1$.

Case 2. If $m = 1$, then

$$\lim_{x \to 0} x^{m-1}e^{-x} = 1.$$

So, the function $f(x) = x^{m-1}e^{-x}$ is bounded in $[0, 1]$. Therefore, the integral $\int_0^1 x^{m-1}e^{-x}dx$ is a proper integral and converges for $m = 1$.

Case 3. If $0 < m < 1$, then

$$\lim_{x \to 0} x^{m-1} e^{-x} = \infty.$$

So, the function $f(x) = x^{m-1} e^{-x}$ is unbounded at $x = 0$.
Let $h(x) = x^{m-1}$.
Then both $f(x)$ and $h(x)$ are unbounded at $x = 0$.

$$\lim_{x \to 0} \frac{f(x)}{h(x)} = \lim_{x \to 0} \frac{x^{m-1} e^{-x}}{x^{m-1}}$$

$$= \lim_{x \to 0} e^{-x} = 1.$$

The integral $\int_0^1 x^{m-1} = \int_0^1 \frac{1}{x^{1-m}}$ converges by the p-test if $p = 1 - m < 1$, that is, $m > 0$. Therefore, by the comparison test, the integral $\int_0^1 x^{m-1} e^{-x} dx$ converges for $0 < m < 1$.

From the above three cases, it is clear that the integral $\int_0^1 x^{m-1} e^{-x} dx$ converges for $m > 0$.

Now, we will check the convergence of $\int_1^\infty x^{m-1} e^{-x} dx$.

Let $f(x) = x^{m-1} e^{-x}$. Then $f(x)$ is bounded in $(1, \infty)$ and

$$\lim_{x \to \infty} x^\mu f(x) = \lim_{x \to \infty} x^\mu \cdot x^{m-1} e^{-x} = \lim_{x \to \infty} x^{\mu+m-1} e^{-x} = \lim_{x \to \infty} \frac{x^{\mu+m-1}}{e^x} 0$$

for all values of x and m (because exponential power always dominates the polynomials).

Taking $\mu > 1$, by the μ-test. the integral $\int_1^\infty x^{m-1} e^{-x} dx$ converges for all values of x and m.

Combining all of the above cases, we conclude that the Gamma-function integral $\int_0^\infty x^{m-1} e^{-x} dx$ converges for $m > 0$.

$\boldsymbol{\mu}$**-test-I.** *Let $f(x)$ be an integrable function in $(a, b]$ and unbounded at $x = a$, and suppose that $\lim_{x \to a+0} (x-a)^\mu f(x) = l$. Then*

a. *The integral $\int_a^b f(x) dx$ converges if $0 < \mu < 1$ and l is a finite number.*

b. *The integral $\int_a^b f(x) dx$ diverges if $\mu \geq 1$ and l is a non-zero number (l can be infinite).*

$\boldsymbol{\mu}$**-test-II.** *Let $f(x)$ be an integrable function in $[a, b)$ and unbounded at $x = b$ and suppose that $\lim_{x \to b-0} (b-x)^\mu f(x) = l'$. Then*

a. *The integral $\int_a^b f(x)dx$ converges if $0 < \mu < 1$ and l' is a finite number.*

b. *The integral $\int_a^b f(x)dx$ diverges if $\mu \geq 1$ and l' is a non-zero number (l' can be infinite).*

Example 1. Discuss the convergence of the integral $\int_1^5 \frac{1}{\sqrt{x^4-1}}dx$.

Solution. The function $f(x) = \frac{1}{\sqrt{x^4-1}}$ is unbounded at $x = 1$. If $\mu = \frac{1}{2}$, then

$$\lim_{x\to 1+} (x-1)^{\frac{1}{2}} \frac{1}{\sqrt{x^4-1}}$$

$$= \lim_{x\to 1+} \sqrt{\frac{x-1}{x^4-1}},$$

$$= \lim_{x\to 1+} \sqrt{\frac{x-1}{(x-1)(x+1)(x^2+1)}}$$

$$= \lim_{x\to 1+} \sqrt{\frac{1}{(x+1)(x^2+1)}} = \frac{1}{2} \text{ (a finite number)}.$$

Since $\mu = \frac{1}{2} < 1$, by the μ-test, the integral $\int_1^5 \frac{1}{\sqrt{x^4-1}}dx$ converges.

Example 2. Discuss the convergence of the integral $\int_0^3 \frac{1}{(3-x)\sqrt{x^2+1}} dx$.

Solution. The function $f(x) = \frac{1}{(3-x)\sqrt{x^2+1}}$ is unbounded at $x = 3$. If $\mu = 1$, then

$$\lim_{x\to 3-0} (3-x)^{\mu} \frac{1}{(3-x)\sqrt{x^2+1}} = \lim_{x\to 3-0} (3-x) \frac{1}{(3-x)\sqrt{x^2+1}},$$

$$= \lim_{x\to 3-0} \frac{1}{\sqrt{x^2+1}} = \frac{1}{\sqrt{10}}.$$

Since $\mu = 1$, by the μ-test, the integral $\int_0^3 \frac{1}{(3-x)\sqrt{x^2+1}}dx$ diverges.

Beta function. The definite integral $\int_0^1 x^{m-1}(1-x)^{n-1}dx$, for $m > 0$ *and* $n > 0$, is known as the Beta function and is denoted by $B(m, n)$. Thus,

$$B(m, n) = \int_0^1 x^{m-1}(1 - x)^{n-1} dx, \quad m > 0, \ n > 0.$$

Example 3 (Convergence of the Beta-function Integral).
Discuss the convergence of the integral $\int_0^1 x^{m-1}(1 - x)^{n-1} dx$.

Solution. The integrand in the above integral is unbounded at both end-points $x = 0$ and $x = 1$, depending on m and n, so the following cases arise:

Case 1. If $m > 1$ and $n > 1$, then

$$\lim_{x \to 0} x^{m-1}(1 - x)^{n-1} = 0 \quad \text{and} \quad \lim_{x \to 1} x^{m-1}(1 - x)^{n-1} = 0.$$

So, the function $f(x) = x^{m-1}(1 - x)^{n-1}$ is bounded in $[0, 1]$. Therefore, the integral $\int_0^1 x^{m-1}(1 - x)^{n-1} dx$ is a proper integral and converges for $m > 1$ and $n > 1$.

Case 2. If $m = 1$ and $n = 1$, then

$$\lim_{x \to 0} x^{m-1}(1 - x)^{n-1} = 1 \quad \text{and} \quad \lim_{x \to 1} x^{m-1}(1 - x)^{n-1} = 1.$$

So, the function $f(x) = x^{m-1}(1 - x)^{n-1}$ is bounded in $[0, 1]$. Therefore, the integral $\int_0^1 x^{m-1}(1 - x)^{n-1} dx$ is a proper integral and converges for $m = 1$ and $n = 1$.

Case 3. If $m < 1$ and $n < 1$, then $f(x)$ is unbounded at both end-points $x = 0$ and 1. We can write

$$\int_0^1 x^{m-1}(1 - x)^{n-1} dx = \int_0^{\frac{1}{2}} x^{m-1}(1 - x)^{n-1} dx$$

$$+ \int_{\frac{1}{2}}^1 x^{m-1}(1 - x)^{n-1} dx.$$

$$= I_1 + I_2.$$

I_1 is unbounded at $x = 0$. First, we will check the convergence of the integral $\int_0^{\frac{1}{2}} x^{m-1}(1 - x)^{n-1} dx$.

$$\lim_{x \to 0+} (x - 0)^{\mu} x^{m-1}(1 - x)^{n-1} = \lim_{x \to 0+} x^{\mu+m-1}(1 - x)^{n-1} = 1,$$

if $\mu + m - 1 = 0$, *that is*, $\mu = 1 - m$. Since $m < 1$, there are two possibilities concerning μ.

If $0 < m < 1$, then $0 < \mu < 1$; and if $m \leq 0$, then $\mu \geq 1$.

So, by the μ-test, I_1 converges if $0 < m < 1$, and diverges if $m \leq 0$.

I_2 is unbounded at $x = 1$. Now, we will check the convergence of the integral $\int_{\frac{1}{2}}^{1} x^{m-1}(1-x)^{n-1}dx$.

$$\lim_{x \to 1-} (1-x)^{\mu} x^{m-1}(1-x)^{n-1} = \lim_{x \to 1-} x^{m-1}(1-x)^{\mu+n-1} = 1,$$

if $\mu + n - 1 = 0$, that is, $\mu = 1 - n$. Since $n < 1$, there are two possibilities concerning μ.

If $0 < n < 1$, then $0 < \mu < 1$; and if $n \leq 0$, then $\mu \geq 1$.

So, by the μ-test, I_2 converges if $0 < n < 1$ and diverges if $n \leq 0$. Therefore, I_1 and I_2 both converge if $0 < m < 1$ and $0 < n < 1$, and both diverge if $m \leq 0$ and $n \leq 0$. From Case 3, we conclude that the integral $\int_0^1 x^{m-1}(1-x)^{n-1}dx$ converges when $0 < m < 1$ and $0 < n < 1$, and diverges when $m \leq 0$ and $n \leq 0$.

Combining all of the above three cases, we conclude that the Beta-function integral $\int_0^1 x^{m-1}(1-x)^{n-1}dx$ converges when $m > 0$ and $n > 0$.

Example 4. Discuss the convergence of the integral $\int_0^1 x^{m-1} \log x\, dx$.

Solution. The integrand in the above integral is unbounded at the end-point $x = 0$ depending on m, so the following cases arise:

Case 1. If $m > 1$, then

$$\lim_{x \to 0+} x^{m-1} \log x = \lim_{x \to 0+} \frac{\log x}{\frac{1}{x^{m-1}}},$$

$$= \lim_{x \to 0+} \frac{1}{(1-m)x^{1-m}} = 0, \quad (\text{since } 1 - m < 0).$$

So, the function $f(x) = x^{m-1} \log x$ is bounded in $[0, 1]$. Therefore, $\int_0^1 x^{m-1} \log x\, dx$ is a proper integral and converges for $m > 1$.

Case 2. If $m = 1$, then

$$\int_0^1 \log x \, dx = \lim_{\varepsilon \to 0} \int_\varepsilon^1 \log x \, dx$$

$$= \lim_{\varepsilon \to 0} [x \log x - x]_\varepsilon^1$$

$$= \lim_{\varepsilon \to 0} [-1 - \varepsilon \log \varepsilon + \varepsilon] = -1.$$

Therefore, the integral $\int_0^1 x^{m-1} \log x \, dx$ converges for $m = 1$.

Case 3. If $m < 1$, then $f(x)$ is unbounded at the end-point $x = 0$. We will check the convergence of the integral $\int_0^1 x^{m-1} \log x \, dx$.

$$\lim_{x \to 0+} (x - 0)^\mu x^{m-1} \log x = \lim_{x \to 0+} x^{\mu+m-1} \log x = 0,$$

if $\mu + m - 1 > 0$, that is, $\mu > 1 - m$.

Since $m < 1$, there are two possibilities concerning μ.

If $0 < m < 1$, then $0 < \mu < 1$; and if $m \le 0$, then $\mu \ge 1$.

So, by the μ-test, the integral $\int_0^1 x^{m-1} \log x \, dx$ converges if $0 < m < 1$ and diverges if $m \le 0$.

Combining all of the above three cases, we conclude that the integral $\int_0^1 x^{m-1} \log x \, dx$ converges when $m > 0$.

Example 5. Discuss the convergence of the integral $\int_0^3 \frac{\log x}{\sqrt[3]{(3-x)}} dx$.

Solution. The integrand $f(x) = \frac{\log x}{\sqrt[3]{(3-x)}}$ is unbounded at both end-points $x = 0$ and $x = 3$. We can write

$$\int_0^3 \frac{\log x}{\sqrt[3]{(3-x)}} dx = \int_0^1 \frac{\log x}{\sqrt[3]{(3-x)}} dx + \int_1^3 \frac{\log x}{\sqrt[3]{(3-x)}} dx$$

$$= I_1 + I_2.$$

$$I_1 = \int_0^1 \frac{\log x}{\sqrt[3]{(3-x)}} dx \quad \text{and} \quad I_2 = \int_1^3 \frac{\log x}{\sqrt[3]{(3-x)}} dx.$$

First. we will check the convergence of I_1. The integrand in I_1 is unbounded at $x = 0$.

$$\lim_{x \to 0+} (x - 0)^\mu \frac{\log x}{\sqrt[3]{(3-x)}} = \lim_{x \to 0+} x^\mu \frac{\log x}{\sqrt[3]{(3-x)}} = 0,$$

if $\mu > 0$, because $\lim_{x \to 0+} x^\mu \log x = 0$.

Choosing $0 < \mu < 1$, by the μ-test, I_1 converges.

Now, we will check convergence of I_2. The integrand in I_2 is unbounded at $x = 3$. If we choose $\mu = \frac{1}{3}$, then

$$\lim_{x \to 3-} (3-x)^{\mu} \frac{\log x}{\sqrt[3]{(3-x)}} = \lim_{x \to 3-} (3-x)^{\frac{1}{3}} \frac{\log x}{\sqrt[3]{(3-x)}}$$

$$= \lim_{x \to 3-} \log x = \log 3.$$

Since $\mu = \frac{1}{3} < 1$, by the μ-test, I_2 converges.

Since I_1 and I_2 both converge, the integral $\int_0^3 \frac{\log x}{\sqrt[3]{(3-x)}} dx$ converges.

Abel's test. Let $h(x)$ be monotonic and bounded in $[a, b]$ and suppose that the integral $\int_a^b f(x)dx$ converges. Then the integral $\int_a^b f(x)h(x)dx$ of their product also converges.

Dirichlet's test. Let $h(x)$ be monotonic and bounded in $[a, b]$ with $\lim_{x \to a} h(x) = 0$ and suppose that $|\int_{a+\delta}^b f(x)dx|$ is bounded for all finite values of x. Then the integral $\int_a^b f(x)h(x)dx$ of their product also converges.

Example 1. Test the convergence of $\int_0^1 \frac{\sin x}{x^2} dx$.

Solution. The integrand $\frac{\sin x}{x^2}$ is unbounded at $x = 0$. Let $f(x) = \sin x$ and $h(x) = \frac{1}{x^2}$. Then $h(x)$ is monotonic and bounded in $[0, 1]$. $\int_0^1 \sin x \, dx$ is a proper integral, so it converges.

By Abel's test, the integral $\int_0^1 \frac{\sin x}{x^2} dx$ converges.

Exercises

1. Test the convergence of the integral $\int_2^\infty \frac{2x^2 dx}{x^4-1}$ and evaluate it if it exists.

   Ans. Convergent, and its value is $\frac{\pi}{2} + \frac{1}{2} \times \log 3 - \tan^{-1} 2$.

2. Test the convergence of the integral $\int_{-\infty}^\infty \frac{dx}{(1+x^2)^2}$ and evaluate it if it exists.

   Ans. Convergent, and its value is $\frac{\pi}{2}$.

3. Test the convergence of the integral $\int_1^\infty \frac{dx}{x}$ and evaluate it if it exists.

Ans. Divergent, i.e. does not exist.

4. Test the convergence of the integral $\int_0^\infty x^3 e^{-x^2}$ and evaluate it if it exists.

Ans. Convergent, and its value is $\frac{1}{2}$.

5. Test the convergence of the integral $\int_1^\infty \frac{1}{x\sqrt{1+x^2}}dx$.

Ans. Convergent.

6. Test the convergence of the integral $\int_0^\infty \frac{x^2}{\sqrt{1+x^5}}dx$.

Ans. Divergent.

7. Test the convergence of the integral $\int_1^\infty \frac{\log x}{x^2}dx$.

Ans. Convergent.

8. Test the convergence of the integral $\int_0^\infty \frac{x^{2m}}{1+x^{2n}}dx$, m and n being positive integers.

Ans. Convergent for $n > m$.

9. Test the convergence of the integral $\int_1^\infty \frac{\sin x}{x^p}dx$.

Ans. Convergent for $p > 0$.

10. Show that the integral $\int_1^\infty \frac{\sin x \log x}{x}dx$ is convergent.

11. Show that the integral $\int_0^\infty \frac{\sin x}{x}dx$ is conditionally convergent (that is, convergent, but not absolutely convergent).

12. Show that the integral $\int_2^\infty \frac{\cos x}{\log x}dx$ is conditionally convergent (that is convergent, but not absolutely convergent).

13. Test the convergence of the following integrals:

a. $\int_0^1 \frac{dx}{x^2}$.

b. $\int_0^1 \frac{dx}{\sqrt{1-x}}$.

c. $\int_0^2 \frac{dx}{(2x-x^2)}$.

Ans. a. Divergent.   b. Convergent.   c.   Divergent.

14. Test the convergence of the following integrals:

a. $\int_0^1 \frac{dx}{\sqrt{1-x^3}}$.

b. $\int_0^1 \frac{\log x\, dx}{\sqrt{x}}$.

c. $\int_1^2 \frac{\sqrt{x}\, dx}{\log x}$.

Ans. a. Convergent.   b. Convergent.   c. Divergent.

15. Test the convergence of the integral $\int_0^2 \frac{\log x}{\sqrt{2-x}}dx$.

Chapter 12

Metric Spaces

In previous chapters, we have studied convergence, continuity, and differentiability on the real line. Now, we extend our notion of general spaces by using the concept of the distance function known as metric. Metric measures distance between two objects. Whether the objects are real or complex numbers, the distance between two objects will always be a real number. So, metric is a rule which assigns two objects a unique real number. In mathematical language, the metric is a function whose domain is the Cartesian product of any arbitrary sets and the codomain is a set of real numbers. The concept of metric spaces was formulated in 1906 by Frechet. The first section of this chapter is devoted mainly to the basic definitions and important examples of metric spaces. Then, we discuss open and closed sets in a metric space. In continuation, we discuss convergence and Cauchy sequences in a metric space. Finally, we consider complete metric spaces and Cantor's intersection theorem.

12.1 Metric Space

Let X be a non-empty set and d be a function such that

$$d : X \times X \to \mathbf{R}.$$

If d satisfies the following properties:

1. $d(x, y) \geq 0, \quad \forall x, y \in X,$
2. $d(x, y) = 0$ iff $x = y, \forall x, y \in X,$

3. $d(x, y) = d(y, x), \quad \forall x, y \in X,$
4. $d(x, y) \leq d(x, z) + d(z, y), \quad \forall x, y, z \in X,$

then d is called a metric on X and the ordered pair (X, d) is called a metric space. The first, third, and fourth properties of a metric space are known as positiveness, symmetricity, and triangle inequality, respectively.

Example 1. Let $d : \mathbf{R} \times \mathbf{R} \to \mathbf{R}$ defined by

$$d(x, y) = |x - y|, \quad \forall x, y \in \mathbf{R}.$$

Then d is a metric on $\mathbf{R}$ and $(\mathbf{R}, d)$ is a metric space. This metric is called usual metric on $\mathbf{R}$.

Example 2. Let $d : X \times X \to \mathbf{R}$, defined by

$$d(x, y) = \begin{cases} 1, & \text{if } x \neq y, \\ 0 & \text{if } x = y, \end{cases} \quad \forall x, y \in X.$$

Then d is a metric on X and (X, d) is a metric space. This metric space is called discrete metric space.

Theorem 1. *Let X be any non-empty set and*

$$d : X \times X \to \mathbf{R}.$$

Then d is a metric on X if and only if it satisfies the following properties:

1. $d(x, y) = 0$ *if and only if $x = y$, $\forall x, y \in X$,*
2. $d(x, y) \leq d(x, z) + d(y, z)$, $\forall x, y, z \in X$.

Proof. Let d be a metric on X. Then we will prove the two points as given in Theorem 1 above. Since d is a metric on X, it satisfies the following properties:

1. $d(x, y) \geq 0$, $\forall x, y \in X$,
2. $d(x, y) = 0$ if and only if $x = y$, $\forall x, y \in X$,
3. $d(x, y) = d(y, x)$, $\forall x, y \in X$,
4. $d(x, y) \leq d(x, z) + d(z, y)$, $\forall x, y, z \in X$.

Now, Property 2 is the same as that which we want to prove in the first point. For the second point, by using Property 3 in Property 4, we get

$$d(x, y) \leq d(x, z) + d(y, z), \quad \forall x, y, z \in X.$$

So, the first part is proved.

Next, we will consider the two points as given in Theorem 1. That is,

1. $d(x, y) = 0$ if and only if $x = y$, $\forall x, y \in X$,
2. $d(x, y) \leq d(x, z) + d(y, z)$, $\forall x, y, z \in X$,

and we will prove that d is a metric on X

For Property 1, by taking $y = x$ in the second point, we get

$$d(x, x) \leq d(x, z) + d(x, z), \quad \forall x, z \in X. \tag{12.1}$$

Using the second point $(d(x, x) = 0)$ in Eq. (12.1), we have

$$0 \leq 2d(x, z), \quad \forall x, z \in X.$$

That is, $d(x, z) \geq 0, \forall x, z \in X$, which is the first property of a metric.

The second property is the same as the second point. So, we do not need to prove it. For the third property, put $z = x$ in the second point, so that

$$d(x, y) \leq d(x, x) + d(y, x), \quad \forall x, y \in X. \tag{12.2}$$

Using the second point $(d(x, x) = 0)$ in Eq. (12.2), we get

$$d(x, y) \leq d(y, x), \quad \forall x, y \in X. \tag{12.3}$$

From the second point, we can write

$$d(y, x) \leq d(y, z) + d(x, z), \quad \forall x, y, z \in X. \tag{12.4}$$

Put $z = y$ in Eq. (12.4), so that

$$d(y, x) \leq d(y, y) + d(x, y), \quad \forall x, y \in X. \tag{12.5}$$

Using the second point $(d(y, y) = 0)$ in Eq. (12.5), we get

$$d(y, x) \leq d(x, y), \quad \forall x, y \in X. \tag{12.6}$$

From Eqs. (12.3) and (12.6), we conclude that

$$d(x, y) = d(y, x) \quad \forall x, y \in X. \tag{12.7}$$

This is the third property of a metric. Now, we will prove the triangle inequality. Using Eq. (12.7) in the second point of Theorem 1, we will get the triangle inequality. So, all four properties of a metric have been proved. Therefore, d is a metric on X.

Theorem 2. *In a metric space* (X, d), *it is asserted that*

$$|d(x, z) - d(z, y)| \leq d(x, y) \quad \forall x, y, z \in X.$$

Proof. Let (X, d) be a metric space. Using the triangle inequality for the points x, z, y we get

$$d(x, z) \leq d(x, y) + d(y, z) \quad \forall x, y, z \in X. \tag{12.8}$$

The second property of the metric can be written as follows:

$$d(y, z) = d(z, y) \quad \forall y, z \in X. \tag{12.9}$$

From Eqs. (12.8) and (12.9), we get

$$d(x, z) \leq d(x, y) + d(z, y) \quad \forall x, y, z \in X.$$
$$\Rightarrow \; d(x, z) - d(z, y) \leq d(x, y) \quad \forall x, y, z \in X. \tag{12.10}$$

Using the triangle inequalities for the points z, y, x, we have

$$d(z, y) \leq d(z, x) + d(x, y) \quad \forall x, y, z \in X. \tag{12.11}$$

The second property of the metric is given as follows:

$$d(z, x) = d(x, z) \quad \forall x, z \in X. \tag{12.12}$$

From Eqs. (12.11) and (12.12), we get

$$d(z, y) \leq d(x, z) + d(x, y) \quad \forall x, y, z \in X.$$
$$\Rightarrow \; d(z, y) - d(x, z) \leq d(x, y) \tag{12.13}$$
$$\Rightarrow \; -(d(x, z) - d(z, y)) \leq d(x, y) \quad \forall x, y, z \in X.$$

From Eqs. (12.10) and (12.13), we get

$$|d(x, z) - d(z, y)| \leq d(x, y) \quad \forall x, y, z \in X.$$

Because, if $a \leq x$ and $-a \leq x$, then $|a| \leq x$.

Pseudo metric. Let X be a non-empty set and d be a function

$$d : X \times X \to \mathbf{R}.$$

If d satisfies the following properties:

1. $d(x, y) \geq 0, \ \forall x, y \in X,$
2. $x = y \ \Rightarrow \ d(x, y) = 0, \ \forall x, y \in X,$
3. $d(x, y) = d(y, x), \ \forall x, y \in X,$
4. $d(x, y) \leq d(x, z) + d(z, y), \ \forall x, y, z \in X,$

then d is called a pseudo metric on X and the ordered pair (X, d) is called a pseudo metric space.

Remark. In a metric space and a pseudo metric space, only the second property differs. The second property in a metric space states that if the distance between two points is zero, then they are the same. But, in a pseudo metric space, the distance between two points can be zero without the points being the same.

Theorem 3. *Every metric space is a pseudo metric space, but the converse is not true.*

Proof. Let (X, d) be a metric space. Then, by definition, it will be a pseudo metric space. To show that the converse is not true, we will give an example. Let d be a function $d : \mathbf{R} \times \mathbf{R} \to \mathbf{R}$, defined by

$$d(x, y) = |x^2 - y^2|, \quad \forall x, y \in \mathbf{R}.$$

Then d is a pseudo metric on $\mathbf{R}$ and $(\mathbf{R}, d)$ is a pseudo metric space. To show that it is not a metric space, let $x = 2$ and $y = -2$. Then

$$d(2, -2) = |2^2 - (-2)^2| = |0| = 0.$$
$$\Rightarrow \ d(x, y) = 0 \text{ but } x \neq y.$$

So, $(\mathbf{R}, d)$ is not a metric space.

l_p-**Space** $(p > 0)$. Let s be the collection of all real or complex sequences. l_p is a collection of those sequences of s which are p^{th} summable. That is,

$$l_p = \left\{ x = (x_1, x_2, \ldots, x_i, \ldots) \in s \text{ such that } \sum_{i=0}^{\infty} |x_i|^p < \infty \right\}.$$

l_∞-**space.** Let s be the collection of all real or complex sequences. l_∞ is the collection of all bounded sequences of s. That is,

$$l_\infty = \left\{ x = (x_1, x_2, \ldots, x_i, \ldots) \in s \text{ such that } |x_i| < \infty \forall i \right\}.$$

Minkowski's inequality $(p > 1)$. Let $x = (x_1, x_2, \ldots, x_i, \ldots) \in l_p$ and $y = (y_1, y_2, \ldots, y_i, \ldots) \in l_p$. Then their sum $x + y = (x_1 + y_1, x_2 + y_2, \ldots, x_i + y_i, \ldots) \in l_p$ and

$$\left(\sum_{i=1}^{\infty} |x_i + y_i|^p \right)^{\frac{1}{p}} \leq \left(\sum_{i=1}^{\infty} |x_i|^p \right)^{\frac{1}{p}} + \left(\sum_{i=1}^{\infty} |y_i|^p \right)^{\frac{1}{p}}.$$

Equivalently, we can also write it as follows:
Let $x = (x_1, x_2, \ldots, x_n) \in R^n$ and $y = (y_1, y_2, \ldots, y_n) \in R^n$ then

$$\left(\sum_{i=1}^{n} |x_i + y_i|^2 \right)^{\frac{1}{2}} \leq \left(\sum_{i=1}^{n} |x_i|^2 \right)^{\frac{1}{2}} + \left(\sum_{i=1}^{n} |y_i|^2 \right)^{\frac{1}{2}}.$$

Example 1. Let d be a function $d : \mathbf{R} \times \mathbf{R} \to \mathbf{R}$, defined by

$$d(x, y) = |x - y|, \quad \forall x, y \in \mathbf{R}.$$

Then show that d is a metric on $\mathbf{R}$ and $(\mathbf{R}, d)$ is a metric space.

Solution. To show that d is a metric on $\mathbf{R}$, we will prove the four properties of metric.

1. Since the modulus of a real number is always non-negative.

$$\Rightarrow \quad \forall x, y \in \mathbf{R} \quad |x - y| \geq 0.$$

That is, $d(x, y) = |x - y| \geq 0. \ \forall x, y \in \mathbf{R}.$

2. Let $d(x, y) = 0$. Then we will show that $x = y$.

$$d(x, y) = |x - y| = 0.$$
$$\Rightarrow x - y = 0 \quad (\text{since } |a| = 0 \Rightarrow a = 0).$$
$$\Rightarrow x = y \quad \forall x, y \in \mathbf{R}.$$

To show the converse, let $x = y$. Then we will show that $d(x, y) = 0$.

$$d(x, y) = |x - y| = |y - y| = |0| = 0, \quad \forall x, y \in \mathbf{R}.$$

3. $d(x, y) = |x - y|$ and

$$d(y, x) = |y - x| = |-(x - y)| = |x - y|.$$
$$\Rightarrow d(x, y) = d(y, x), \quad \forall x, y \in \mathbf{R}.$$

4. $d(x, y) = |x - y| = |x - z + z - y|$
$\leq |x - z| + |z - y|$ (since $|a + b| \leq |a| + |b|$, where $a = x - z$ and $b = z - y$).

Using the definition of a metric, $d(x, z) = |x - z|$ and $d(z, y) = |z - y|$.

$$\Rightarrow d(x, y) \leq d(x, z) + d(z, y), \quad \forall x, y, z \in \mathbf{R}.$$

Since all properties of the metric are satisfied, d is a metric on $\mathbf{R}$ and $(\mathbf{R}, d)$ is a metric space.

Example 2. $d : X \times X \rightarrow \mathbf{R}$, defined by

$$d(x, y) = \begin{cases} 1, & \text{if } x \neq y, \\ 0 & \text{if } x = y, \end{cases} \quad \forall x, y \in X.$$

Then show that d is a metric on X and (X, d) is a metric space.

Solution. To show that d is a metric on X, we will prove the four properties of a metric.

1. Since there are only two possibilities, either $d(x, y) = 1$ or $d(x, y) = 0$, therefore,

$$d(x, y) \geq 0, \quad \forall x, y \in X.$$

2. Let $d(x, y) = 0$. Then we will show that $x = y$.

By the definition of a metric, it is clear that $d(x, y) = 0$.

$$\Rightarrow \quad x = y, \quad \forall x, y \in X.$$

To show the converse, let $x = y$. Then we will show that $d(x, y) = 0$. Again, by the definition of a metric, it is clear that

$$d(x, y) = 0 \quad \text{if } x = y \quad \forall x, y \in X.$$

3. If $x \neq y$ then $d(x, y) = 1$ and $d(y, x) = 1$.

If $x = y$ then $d(x, y) = 0$ and $d(y, x) = 0$.
Therefore, $d(x, y) = d(y, x), \forall x, y \in X$.

4. To show the triangle inequality, the following five cases arise:

Case 1. If $x = y = z$, then

$$d(x, y) = 0, \quad d(x, z) = 0 \quad \text{and} \quad d(z, y) = 0.$$

Now, we have $d(x, y) = d(x, z) + d(z, y), \forall x, y, z \in \mathbf{R}$.
We can also write $d(x, y) \leq d(x, z) + d(z, y), \forall x, y, z \in \mathbf{R}$.

Case 2. If $x \neq y \neq z$, then

$$d(x, y) = 1, \quad d(x, z) = 1 \quad \text{and} \quad d(z, y) = 1.$$

Now, we have $d(x, z) + d(z, y) = 1 + 1 = 2, \forall x, y, z \in \mathbf{R}$.

$$\Rightarrow d(x, y) \leq d(x, z) + d(z, y), \quad \forall x, y, z \in \mathbf{R}.$$

Case 3. If $x = y \neq z$, then

$$d(x, y) = 0, \quad d(x, z) = 1 \quad \text{and} \quad d(z, y) = 1.$$

Now, we have $d(x, z) + d(z, y) = 1 + 1 = 2, \forall x, y, z \in \mathbf{R}$.

$$\Rightarrow d(x, y) \leq d(x, z) + d(z, y), \quad \forall x, y, z \in \mathbf{R}.$$

Case 4. If $x \neq y = z$, then

$$d(x, y) = 1, \quad d(x, z) = 1 \quad \text{and} \quad d(z, y) = 0.$$

Now, we have $d(x, z) + d(z, y) = 1 + 0 = 1$, $\forall x, y, z \in \mathbf{R}$.

$\Rightarrow d(x, y) \le d(x, z) + d(z, y)$, $\quad \forall x, y, z \in \mathbf{R}$.

Case 5. If $x = z \neq y$, then

$$d(x, y) = 1, \quad d(x, z) = 0, \quad \text{and} \quad d(z, y) = 1.$$

Now, we have $d(x, z) + d(z, y) = 0 + 1 = 1$, $\forall x, y, z \in \mathbf{R}$.

$\Rightarrow d(x, y) \le d(x, z) + d(z, y)$, $\quad \forall x, y, z \in \mathbf{R}$.

In all cases, therefore, triangle inequality is proved.

Since all properties of the metric are satisfied, d is a metric on X and (X, d) is a metric space.

Example 3. Let d be a function $d : \mathbf{R}^2 \times \mathbf{R}^2 \to \mathbf{R}$, defined by

$$d(x, y) = \sqrt{(x_1 - y_1)^2 + (x_2 - y_2)^2}, \quad \forall x = (x_1, x_2), y = (y_1, y_2) \in \mathbf{R}^2.$$

Then show that d is a metric on $\mathbf{R}^2$ and $(\mathbf{R}^2, d)$ is a metric space.

Solution. To show that d is a metric on $\mathbf{R}$, we will prove four properties of metric.

The square of a real number is always non-negative.

$$\Rightarrow (x_1 - y_1)^2 + (x_2 - y_2)^2 \ge 0, \quad \forall x, y \in \mathbf{R}^2.$$

$$\Rightarrow \sqrt{(x_1 - y_1)^2 + (x_2 - y_2)^2} \ge 0, \quad \forall x, y \in \mathbf{R}^2.$$

$$\Rightarrow d(x, y) \ge 0, \quad \forall x, y \in \mathbf{R}^2.$$

1. Let $d(x, y) = 0$. Then we will show that $x = y$.

$$d(x, y) = \sqrt{(x_1 - y_1)^2 + (x_2 - y_2)^2} = 0.$$

$$\Rightarrow (x_1 - y_1)^2 + (x_2 - y_2)^2 = 0, \quad (\text{since } \sqrt{a} = 0 \Rightarrow a = 0).$$

$$\Rightarrow (x_1 - y_1)^2 = 0 \quad \text{and} \quad (x_2 - y_2)^2 = 0.$$

$$\Rightarrow x_1 - y_1 = 0 \quad \text{and} \quad x_2 - y_2 = 0.$$

$$\Rightarrow x_1 = y_1 \quad \text{and} \quad x_2 = y_2.$$

$$\Rightarrow x = y, \quad \forall x, y \in \mathbf{R}^2.$$

Similarly, from the reverse process, we get the converse part.

2. $d(x, y) = \sqrt{(x_1 - y_1)^2 + (x_2 - y_2)^2}$ and

$$d(y, x) = \sqrt{(y_1 - x_1)^2 + (y_2 - x_2)^2} = \sqrt{(x_1 - y_1)^2 + (x_2 - y_2)^2}.$$

$$\Rightarrow \ d(x, y) = d(y, x), \quad \forall x, y \in \mathbf{R}^2.$$

3. Let $x = (x_1, x_2)$, $y = (y_1, y_2)$, $z = (z_1, z_2) \in \mathbf{R}^2$.

$$d(x, y) = \sqrt{(x_1 - y_1)^2 + (x_2 - y_2)^2}$$

$$= \sqrt{\sum_{i=1}^{2} (x_i - y_i)^2}$$

$$= \sqrt{\sum_{i=1}^{2} |x_i - y_i|^2} \quad (\text{since } (a)^2 = |a|^2)$$

$$= \sqrt{\sum_{i=1}^{2} |x_i - z_i + z_i - y_i|^2}.$$

Let $a_i = x_i - z_i$ and $b_i = z_i - y_i$. Then

$$d(x, y) = \sqrt{\sum_{i=1}^{2} |a_i + b_i|^2},$$

$$\leq \left(\sum_{i=1}^{2} |a_i|^2 \right)^{\frac{1}{2}} + \left(\sum_{i=1}^{2} |b_i|^2 \right)^{\frac{1}{2}},$$

(by Minkowski's inequality).

Putting $a_i = x_i - z_i$ and $b_i = z_i - y_i$, we have

$$d(x, y) \leq \left(\sum_{i=1}^{2} |x_i - z_i|^2 \right)^{\frac{1}{2}} + \left(\sum_{i=1}^{2} |z_i - y_i|^2 \right)^{\frac{1}{2}},$$

$$\Rightarrow \ d(x, y) \leq \sqrt{|x_1 - z_1|^2 + |x_2 - z_2|^2}$$

$$+ \sqrt{|z_1 - y_1|^2 + |z_2 - y_2|^2},$$

$$\Rightarrow d(x,y) \le \sqrt{(x_1 - z_1)^2 + (x_2 - z_2)^2}$$

$$+ \sqrt{(z_1 - y_1)^2 + (z_2 - y_2)^2}.$$

Using the definition of a metric, we get

$$d(x,z) = \sqrt{(x_1 - z_1)^2 + (x_2 - z_2)^2} \quad \text{and}$$

$$d(z,y) = \sqrt{(z_1 - y_1)^2 + (z_2 - y_2)^2}.$$

$$\Rightarrow d(x,y) \le d(x,z) + d(z,y), \quad \forall x,y,z \in \mathbf{R}^2.$$

Since all properties of the metric are satisfied, d is a metric on $\mathbf{R}^2$ and $(\mathbf{R}^2, d)$ is a metric space.

Example 4. Let d be a function $d : \mathbf{R} \times \mathbf{R} \to \mathbf{R}$ defined by

$$d(x,y) = \frac{|x - y|}{1 + |x - y|}, \quad \forall x, y \in \mathbf{R}.$$

Then show that d is a metric on $\mathbf{R}$ and $(\mathbf{R}, d)$ is a metric space.

Solution. To show that d is a metric on $\mathbf{R}$, we will prove the four properties of a metric.

The modulus of a real number is always non-negative.

$$\Rightarrow \forall x, y \in \mathbf{R}, \quad |x - y| \ge 0.$$

$$\Rightarrow \forall x, y \in \mathbf{R}, \quad \frac{|x - y|}{1 + |x - y|} \ge 0.$$

That is, $d(x,y) = \frac{|x-y|}{1+|x-y|} \ge 0, \forall x, y \in \mathbf{R}.$

1. Let $d(x,y) = 0$. Then we will show that $x = y$.

$$\Rightarrow d(x,y) = \frac{|x - y|}{1 + |x - y|} = 0.$$

$$\Rightarrow |x - y| = 0.$$

$$\Rightarrow x - y = 0.$$

$$\Rightarrow x = y, \quad \forall x, y \in \mathbf{R}.$$

To show the converse, let $x = y$. We will show that $d(x, y) = 0$.

$$d(x, y) = \frac{|x - y|}{1 + |x - y|} = \frac{|y - y|}{1 + |y - y|} = |\frac{0}{1}| = 0, \quad \forall x, y \in \mathbf{R}.$$

2. $d(x, y) = \frac{|x-y|}{1+|x-y|}$ and

$$d(y, x) = \frac{|y - x|}{1 + |y - x|} = \frac{|-(x - y)|}{1 + |-(x - y)|} = \frac{|x - y|}{1 + |x - y|}.$$

$$\Rightarrow d(x, y) = d(y, x), \quad \forall x, y \in \mathbf{R}.$$

3. $|x - y| = |x - z + z - y|$

$\leq |x - z| + |z - y|$ (since $|a + b| \leq |a| + |b|$,

where $a = x - z$ and $b = z - y$).

$$\Rightarrow \frac{|x - y|}{1 + |x - y|} \leq \frac{|x - z| + |z - y|}{1 + |x - z| + |z - y|}$$

$$\left(\text{since, if } |a| \leq |b|, \text{ then } \frac{|a|}{1 + |a|} \leq \frac{|b|}{1 + |b|} \right).$$

$$\Rightarrow \frac{|x - y|}{1 + |x - y|} \leq \frac{|x - z|}{1 + |x - z| + |z - y|} + \frac{|z - y|}{1 + |x - z| + |z - y|}.$$

$$\Rightarrow \frac{|x - y|}{1 + |x - y|} \leq \frac{|x - z|}{1 + |x - z|} + \frac{|z - y|}{1 + |z - y|}.$$

Using the definition of a metric,

$$d(x, y) = \frac{|x - y|}{1 + |x - y|}, \quad d(x, z) = \frac{|x - z|}{1 + |x - z|}$$

$$\text{and} \quad d(z, y) = \frac{|z - y|}{1 + |z - y|}.$$

$$\Rightarrow d(x, y) \leq d(x, z) + d(z, y), \quad \forall x, y, z \in \mathbf{R}.$$

Since all properties of the metric are satisfied, d is a metric on $\mathbf{R}$ and $(\mathbf{R}, d)$ is a metric space.

Example 5. Let d be a metric on $\mathbf{R}$ and define a function $d_1 \colon \mathbf{R} \times \mathbf{R} \to \mathbf{R}$ by

$$d_1(x, y) = \frac{d(x, y)}{a + d(x, y)}, \quad \text{where } a \geq 1 \quad \text{and} \quad \forall x, y \in \mathbf{R}.$$

Then show that d_1 is a metric on $\mathbf{R}$ and $(\mathbf{R}, d_1)$ is a metric space.

Solution. It is given that d is a metric on $\mathbf{R}$. To show that d_1 is a metric on $\mathbf{R}$, we will prove the four properties of the metric.

1. Since d is a metric on $\mathbf{R}$

$$d(x,y) \geq 0, \quad \forall x, y \in \mathbf{R}.$$

$$\Rightarrow \frac{d(x,y)}{a + d(x,y)} \geq 0, \quad \text{for } a \geq 1 \text{ and } \forall x, y \in \mathbf{R}.$$

$$\Rightarrow d_1(x,y) = \frac{d(x,y)}{a + d(x,y)} \geq 0, \quad \text{for } a \geq 1 \text{ and } \forall x, y \in \mathbf{R}.$$

2. Let $d_1(x,y) = 0$. Then we will show that $x = y$.

$$\Rightarrow d_1(x,y) = \frac{d(x,y)}{a + d(x,y)} = 0.$$

$$\Rightarrow d(x,y) = 0.$$

Since d is a metric, $d(x,y) = 0$,

$$\Rightarrow x = y, \forall x, y \in \mathbf{R}.$$

To show the converse, let $x = y$. Then we will show that $d_1(x,y) = 0$.
Since d is a metric and $x = y$, therefore, $d(x,y) = 0$.

$$\Rightarrow \quad d_1(x,y) = \frac{d(x,y)}{a + d(x,y)} = \left| \frac{0}{a + 0} \right| = 0, \quad \forall x, y \in \mathbf{R}.$$

3. $d_1(x,y) = \frac{d(x,y)}{a + d(x,y)}$ and

$$d_1(y,x) = \frac{d(y,x)}{a + d(y,x)}.$$

Since d is a metric,

$$d(x,y) = d(y,x), \quad \forall x, y \in \mathbf{R}.$$

$$\Rightarrow d_1(y,x) = \frac{d(y,x)}{a + d(y,x)} = \frac{d(x,y)}{a + d(x,y)}, \quad \forall x, y \in \mathbf{R}.$$

$$\Rightarrow d_1(x,y) = d_1(y,x), \quad \forall x, y \in \mathbf{R}.$$

4. Since d is a metric, by the triangle inequality, we have

$$d(x, y) \leq d(x, z) + d(z, y), \quad \forall x, y, z \in \mathbf{R}. \tag{12.14}$$

If $c, d \geq 0$ with $c \leq d$ and $a \geq 1$, then

$$\frac{c}{a + c} \leq \frac{d}{a + d}, \tag{12.15}$$

From Eqs. (12.14) and (12.15), we can write

$$\frac{d(x, y)}{a + d(x, y)} \leq \frac{d(x, z) + d(z, y)}{a + d(x, z) + d(z, y)},$$

$$\Rightarrow \frac{d(x, y)}{a + d(x, y)} \leq \frac{d(x, z)}{a + d(x, z) + d(z, y)} + \frac{d(z, y)}{a + d(x, z) + d(z, y)}.$$

Since d is a metric, we have $d(x, z)$ and $d(z, y) \geq 0$.

$$\Rightarrow \frac{d(x, y)}{a + d(x, y)} \leq \frac{d(x, z)}{a + d(x, z)} + \frac{d(z, y)}{a + d(z, y)}.$$

Using the definition of a metric, we get

$$d_1(x, y) = \frac{d(x, y)}{a + d(x, y)}, d_1(x, z) = \frac{d(x, z)}{a + d(x, z)},$$

$$\text{and} \quad d_1(z, y) = \frac{d(z, y)}{a + d(z, y)}.$$

$$\Rightarrow d_1(x, y) \leq d_1(x, z) + d_1(z, y), \quad \forall x, y, z \in \mathbf{R}.$$

Since all properties of the metric are satisfied, d_1 is a metric on $\mathbf{R}$ and $(\mathbf{R}, d_1)$ is a metric space.

12.2   Neighborhoods and Open Spheres

Let (X, d) be a metric space and $a \in X$. Then the neighborhood of a is the set $\{x \in X : d(x, a) < r\}$, where $r > 0$. It is denoted by $N(a, r)$ and r is called its radius.

It is also known as open sphere with the center at a and radius $r > 0$ and is denoted by $S(a, r)$. It is defined as

$$S(a, r) = \{y \in X : d(y, a) < r\}.$$

Closed sphere. A closed sphere with center at a and radius $r > 0$ is denoted as $S[a, r]$ and it is defined as follows:

$$S[a, r] = \{y \in X : d(y, a) \leq r\}.$$

Remark. In an open sphere, the boundary is not included ($<$), but in a closed sphere, the boundary is included ($\leq$).

Example 1. Let $X = [0, 1]$ and suppose that d is a metric $d :$ $X \times X \to \mathbf{R}$, defined by $d(x, y) = |x - y|$, $\forall x, y \in X$. Then find

a. $S\left(\frac{1}{2}, \frac{1}{3}\right)$ and $S\left[\frac{1}{2}, \frac{1}{3}\right]$.
b. $S\left(1, \frac{1}{2}\right)$ and $S\left[1, \frac{1}{2}\right]$.
c. $S\left(\frac{1}{4}, 1\right)$ and $S\left[\frac{1}{4}, 1\right]$.

Solution. a. In $S\left(\frac{1}{2}, \frac{1}{3}\right)$, $a = \frac{1}{2}$ and $r = \frac{1}{3}$.

$$S\left(\frac{1}{2}, \frac{1}{3}\right) = \left\{y \in X : d\left(y, \frac{1}{2}\right) < \frac{1}{3}\right\}.$$

$$= \left\{y \in [0, 1] : \left|y - \frac{1}{2}\right| < \frac{1}{3}\right\}.$$

If $|a| < b$ then $a < b$ and $-a < b$. That is,

$$\left(y - \frac{1}{2}\right) < \frac{1}{3} \quad \text{and} \quad -\left(y - \frac{1}{2}\right) < \frac{1}{3},$$

$$\Rightarrow y < \frac{5}{6} \quad \text{and} \quad y > \frac{1}{6}.$$

$$\Rightarrow S\left(\frac{1}{2}, \frac{1}{3}\right) = \left\{y \in [0, 1] : \frac{1}{6} < y < \frac{5}{6}\right\},$$

$$\Rightarrow S\left(\frac{1}{2}, \frac{1}{3}\right) = \left(\frac{1}{6}, \frac{5}{6}\right).$$

For the closed sphere given by

$$S\left[\frac{1}{2}, \frac{1}{3}\right] = \left\{y \in X : d(y, \frac{1}{2}) \leq \frac{1}{3}\right\},$$

$$= \left\{y \in [0,1] : \left|y - \frac{1}{2}\right| \leq \frac{1}{3}\right\}.$$

If $|a| \leq b$ then $a \leq b$ and $-a \leq b$. That is,

$$\left(y - \frac{1}{2}\right) \leq \frac{1}{3} \quad \text{and} \quad -\left(y - \frac{1}{2}\right) \leq \frac{1}{3},$$

$$\Rightarrow y \leq \frac{5}{6} \quad \text{and} \quad y \geq \frac{1}{6}.$$

$$\Rightarrow S\left[\frac{1}{2}, \frac{1}{3}\right] = \left\{y \in [0,1] : \frac{1}{6} \leq y \leq \frac{5}{6}\right\},$$

$$\Rightarrow S\left[\frac{1}{2}, \frac{1}{3}\right] = \left[\frac{1}{6}, \frac{5}{6}\right].$$

b. In $S\left(1, \frac{1}{2}\right)$, $a = 1$ and $r = \frac{1}{2}$.

$$S\left(1, \frac{1}{2}\right) = \left\{y \in X : d(y, 1) < \frac{1}{2}\right\}.$$

$$= \left\{y \in [0,1] : |y - 1| < \frac{1}{2}\right\}.$$

Solving the inequality, we get

$$(y - 1) < \frac{1}{2} \quad \text{and} \quad -(y - 1) < \frac{1}{2},$$

$$\Rightarrow y < \frac{3}{2} \quad \text{and} \quad y > \frac{1}{2}.$$

$$\Rightarrow S\left(1, \frac{1}{2}\right) = \left\{y \in [0,1] : \frac{1}{2} < y < \frac{3}{2}\right\},$$

$$\Rightarrow S\left(1, \frac{1}{2}\right) = \left(\frac{1}{2}, 1\right].$$

For the closed sphere given by

$$S\left[1, \frac{1}{2}\right] = \left\{ y \in X : d(y, 1) \leq \frac{1}{2} \right\},$$

$$= \left\{ y \in [0, 1] : |y - 1| \leq \frac{1}{2} \right\}.$$

Solving the inequality, we get

$$(y - 1) \leq \frac{1}{2} \quad \text{and} \quad -(y - 1) \leq \frac{1}{2},$$

$$\Rightarrow y \leq \frac{3}{2} \quad \text{and} \quad y \geq \frac{1}{2}.$$

$$\Rightarrow S\left[1, \frac{1}{2}\right] = \left\{ y \in [0, 1] : \frac{1}{2} \leq y \leq \frac{3}{2} \right\},$$

$$\Rightarrow S\left[1, \frac{1}{2}\right] = \left[\frac{1}{2}, 1\right].$$

c. In $S\left(\frac{1}{4}, 1\right)$, $a = \frac{1}{4}$ and $r = 1$.

$$S\left(\frac{1}{4}, 1\right) = \left\{ y \in X : d\left(y, \frac{1}{4}\right) < 1 \right\}.$$

$$= \left\{ y \in [0, 1] : |y - \frac{1}{4}| < 1 \right\}.$$

Solving the inequality, we get

$$\left(y - \frac{1}{4}\right) < 1 \quad \text{and} \quad -\left(y - \frac{1}{4}\right) < 1,$$

$$\Rightarrow y < \frac{5}{4} \quad \text{and} \quad y > -\frac{3}{4}.$$

$$\Rightarrow S\left(\frac{1}{4}, 1\right) = \left\{ y \in [0, 1] : -\frac{3}{4} < y < \frac{5}{4} \right\},$$

$$\Rightarrow S\left(\frac{1}{4}, 1\right) = (0, 1).$$

For the closed sphere given by

$$S\left[\frac{1}{4}, 1\right] = \left\{ y \in X : d(y, \frac{1}{4}) \leq 1 \right\},$$

$$= \left\{ y \in [0, 1] : |y - \frac{1}{4}| \leq 1 \right\}.$$

Solving the inequality, we get

$$\left(y - \frac{1}{4} \right) \leq 1 \quad \text{and} \quad - \left(y - \frac{1}{4} \right) \leq 1,$$

$$\Rightarrow y \leq \frac{5}{4} \quad \text{and} \quad y \geq -\frac{3}{4}.$$

$$\Rightarrow S\left[\frac{1}{2}, \frac{1}{3}\right] = \left\{ y \in [0, 1] : -\frac{3}{4} \leq y \leq \frac{5}{4} \right\},$$

$$\Rightarrow S\left[\frac{1}{2}, \frac{1}{3}\right] = [0, 1].$$

Example 2. Discuss open and closed spheres in **R** with respect to the usual metric d defined by

$$d(x, y) = |x - y|, \quad \forall x, y \in \mathbf{R}.$$

Solution. $(\mathbf{R}, d)$ is a metric space. Let $r > 0$. Then the open sphere is given by

$$S(a, r) = \{ y \in \mathbf{R} : d(y, a) < r \},$$
$$= \{ y \in \mathbf{R} : |y - a| < r \},$$
$$= \{ y \in \mathbf{R} : a - r < y < a + r \}.$$

Using the definition of open intervals, we get

$$S(a, r) = (a - r \, a + r).$$

Therefore, open spheres on **R** with respect to usual metric are open intervals.

Now, for a closed sphere, we have

$$S[a, r] = \{y \in \mathbf{R} : d(y, a) \leq r\},$$
$$= \{y \in \mathbf{R} : |y - a| \leq r\},$$
$$= \{y \in \mathbf{R} : a - r \leq y \leq a + r\}.$$

Using the definition of closed intervals, we get

$$S(a, r) = [a - r \, a + r].$$

Therefore, closed spheres on $\mathbf{R}$ with respect to the usual metric are closed intervals.

Example 3. Discuss open spheres in $\mathbf{R}^2$ with respect to the following metrics:

a. $d_1(x, y) = \sqrt{(x_1 - y_1)^2 + (x_2 - y_2)^2}$, $\forall x = (x_1, x_2), y = (y_1, y_2) \in \mathbf{R}^2$,

b. $d_2(x, y) = |x_1 - y_1| + |x_2 - y_2|$, $\forall x = (x_1, x_2), y = (y_1, y_2) \in \mathbf{R}^2$.

Also find $S((0, 0), 2)$.

Solution. a. $(\mathbf{R}^2, d_1)$ is a metric space. Let $y = (y_1, y_2), a = (a_1, a_2) \in \mathbf{R}^2$ and $r > 0$. Then

$$S(a, r) = \{y = (y_1, y_2) \in \mathbf{R}^2 : d_1(y, a) < r\},$$

$$= \{y = (y_1, y_2) \in \mathbf{R}^2 : \sqrt{(y_1 - a_1)^2 + (y_2 - a_2)^2} < r\},$$

$$= \{y = (y_1, y_2) \in \mathbf{R}^2 : (y_1 - a_1)^2 + (y_2 - a_2)^2 < r^2\}.$$

So open spheres are the interior of the circles with center $a = (a_1, a_2)$ and radius r.

For $S((0, 0), 2)$, taking $a = (0, 0)$ and $r = 2$.

$$S((0, 0), 2) = \{y = (y_1, y_2) \in \mathbf{R}^2 : (y_1 - 0)^2 + (y_2 - 0)^2 < 2^2\}.$$

Therefore, $S((0, 0), 2)$ is the interior of the circle $y_1^2 + y_2^2 = 4$.

b. $(\mathbf{R}^2, d_2)$ is a metric space. Let $y = (y_1, y_2), a = (a_1, a_2) \in \mathbf{R}^2$, and $r > 0$. Then

$$S(a, r) = \{y = (y_1, y_2) \in \mathbf{R}^2 : d_2(y, a) < r\},$$

$$= \{y = (y_1, y_2) \in \mathbf{R}^2 : |y_1 - a_1| + |y_2 - a_2| < r\}.$$

So, open spheres are the interior of the squares.

For $S((0,0),2)$ taking $a = (0,0)$ and $r = 2$, we have

$$S((0,0),2) = \{y = (y_1, y_2) \in \mathbf{R}^2 : |y_1 - 0| + |y_2 - 0| < 2\}.$$

$$\Rightarrow\ S((0,0),2) = \{y = (y_1, y_2) \in \mathbf{R}^2 : |y_1| + |y_2| < 2\}.$$

Therefore, $S((0,0),2)$ is the interior of the square $|y_1| + |y_2| < 2$ or square formed by the coordinates $(-2,0), (0,-2), (2,0)$, and $(0,2)$.

Note. From the above discussion, it is clear that nature of the open spheres depends on the metric defined on the set. On the same set, open spheres can be circles or squares depending on the metric defined on that set.

Interior point. Let (X, d) be a metric space and $G \subseteq X$. A point $a \in G$ is called an interior point of G if there exists an open sphere with center a and contained in G. That is, for some $r > 0$,

$$a \in S(a,r) \subseteq G.$$

Interior of a set. Let (X, d) be a metric space and $G \subseteq X$. Then the interior of a set G is the collection of all interior points of G and it is denoted by int G.

Limit point. Let (X, d) be a metric space and $F \subseteq X$. A point $a \in X$ is called a limit point of F if every **nbd** of a contains a point of F other than a. Let $N(a,r)$ be an arbitrary **nbd**. Then

$$N(a,r) - \{a\} \bigcap F \neq \varphi.$$

Derived set. Let (X, d) be a metric space and $F \subseteq X$. Then the derived set of a set F is the collection of all limit points of F and it is denoted by der F or F'.

Open set. Let (X, d) be a metric space and $G \subseteq X$. Then G is called an open set if all of its points are interior points. That is, if $a \in G$ is an arbitrary point, then there exists an open sphere with center at a and contained in G. That is, for some $r > 0$,

$$a \in S(a,r) \subseteq G.$$

Example 1. Let $X = \mathbf{R}$ and let d be a metric $d : \mathbf{R} \times \mathbf{R} \to \mathbf{R}$, defined by

$$d(x,y) = |x - y|, \quad \forall x, y \in \mathbf{R}.$$

Then $(0,1)$ is an open set in the metric space $(\mathbf{R}, d)$.

Example 2. $d : X \times X \to \mathbf{R}$, defined by

$$d(x, y) = \begin{cases} 1, & \text{if } x \neq y, \\ 0 & \text{if } x = y, \end{cases} \quad \forall x, y \in X.$$

Then (X, d) is a discrete metric space. Discuss open spheres in this space and show that every singleton set is an open set.

Solution. Let $a \in X$. Taking $0 < r \leq 1$, we have

$$S(a, r) = \{y \in X : d(y, a) < r\},$$
$$= \{a\} \text{ (because } d(a, a) = 0 \text{ and } d(y, a) = 1 \not< r, \text{ if } y \neq a).$$
$$\Rightarrow S(a, r) \subseteq G = \{a\}.$$

If $r > 1$, then

$$S(a, r) = \{y \in X : d(y, a) < r\},$$
$$= X \text{ (because } d(a, a) = 0 \text{ and } d(y, a) = 1 < r, \text{ if } y \neq a).$$

In this case, $S(a, r) \nsubseteq G$.

Therefore, in a discrete metric space, open spheres are singleton sets and whole set.

If we take any number r with $0 < r \leq 1$, then the condition of open sets is satisfied. In particular, we have

$$S\left(a, \frac{1}{3}\right) = \{a \subseteq G = \{a\}.$$

Therefore, $G = \{a\}$ is an open set.

Adherent points. Let (X, d) be a metric space and let $F \subseteq X$. Then a number $a \in X$ is called an adherent point of F. If we take any open sphere centered at a then it contains a point of F. That is, for every $r > 0$, we have

$$F \cap S(a, r) \neq \Phi.$$

So, from the definition it is clear that any point of the set, that is, $a \in F$ is an adherent point, because, if $a \in F$, then $F \cap S(a, r) \neq \Phi$, and it contains at least one point a always. By the definition of limit points, it is clear that each limit point is an adherent point.

Closure of a set. Let (X, d) be a metric space and let $F \subseteq X$. Then the closure of the set F is the collection of all adherent points of F and it is denoted by $\tilde{F}$.

$$\tilde{F} = F \cup F'.$$

Closed set. Let (X, d) be a metric space and $F \subseteq X$. Then F is called a closed set if it has all of its limit points. That is, no point outside from F can be a limit point of F. We can also say that a set F is closed if $\tilde{F} = F$.

Thus, a set F is closed if and only if $F' \subseteq F$.

Example 1. Let $X = \mathbf{R}$ and let d be a metric $d : \mathbf{R} \times \mathbf{R} \to \mathbf{R}$, defined by

$$d(x, y) = |x - y|, \quad \forall x, y \in \mathbf{R}.$$

Then $[0, 1]$ is a closed set in the metric space $(\mathbf{R}, d)$.

Theorem 1. *In a metric space, every open sphere is an open set.*

Proof. Let (X, d) be a metric space and let $S(a, r)$ be an open sphere. Then we will show that $S(a, r)$ is an open set. For this, we need to show that each point in $S(a, r)$ is an interior point.

Let $y \in S(a, r)$ be an arbitrary point. Then we will show that y is an interior point. For this, we need to show that there exists an open sphere centered at y and contained in $S(a, r)$.

Since $y \in S(a, r)$ by the definition of open sphere, we have

$$d(y, a) < r.$$
$$\Rightarrow \ r - d(y, a) > 0.$$

Let $r_1 = r - d(y, a) > 0$.

$$\Rightarrow \quad d(y, a) = r - r_1. \tag{12.16}$$

Now, we will show that the open sphere $S(y, r_1)$ is contained in $S(a, r)$.

Let $x \in S(y, r_1)$ be an arbitrary point. Then we will show that $x \in S(a, r)$ for which we will prove that $d(x, a) < r$.

Since $x \in S(y, r_1)$

$$\Rightarrow \quad d(x, y) < r_1, \tag{12.17}$$

(X, d) is a metric space. Using the triangle inequality for points x, a and y,

$$d(x, a) \leq d(x, y) + d(y, a).$$

Using Eq. (12.17) in the above inequality, we obtain

$$d(x, a) < r_1 + d(y, a).$$

Using Eq. (12.16), we get

$$d(x, a) < r_1 + r - r_1.$$
$$\Rightarrow d(x, a) < r.$$
$$\Rightarrow x \in S(a, r).$$

That is, x is an arbitrary point. So, each point in $S(y, r_1)$ is also in $S(a, r)$.

$$\Rightarrow S(y, r_1) \subseteq S(a, r).$$

So, y is an interior point. Since y is an arbitrary point, each point in $S(a, r)$ is an interior point.

Therefore, $S(a, r)$ is an open set.

Theorem 2. *In a metric space, any arbitrary union of open sets is an open set.*

Proof. Let (X, d) be a metric space. Let $\{A_\lambda\}_{\lambda \in \Lambda}$ be an arbitrary collection of open sets. Then we will prove that $A = \bigcup_{\lambda \in \Lambda} A_\lambda$ is an open set.

Let $x \in A = \bigcup_{\lambda \in \Lambda} A_\lambda$. Then we will show that there exists an open sphere containing x and contained in A.

Since $x \in A = \bigcup_{\lambda \in \Lambda} A_\lambda$

$\Rightarrow \quad x \in A_\lambda$, for at least one $\lambda \in \Lambda$.

Since A_λ is an open set, there exists an open sphere $S(x,r)$ such that

$$x \in S(x,r) \subseteq A_\lambda,$$

$$\Rightarrow x \in S(x,r) \subseteq A_\lambda \subseteq \bigcup_{\lambda \in \wedge} A_\lambda.$$

$$\Rightarrow \quad x \in S(x,r) \subseteq A = \bigcup_{\lambda \in \wedge} A_\lambda.$$

Since x is an arbitrary point, each point of A is an interior point and, therefore, $A = \bigcup_{\lambda \in \wedge} A_\lambda$ is an open set.

Theorem 3. *In a metric space, finite intersection of open sets is an open set.*

Proof. Let (X,d) be a metric space. Let $\{A_\lambda\}_{\lambda \in \wedge}$, where $\wedge = \{1, 2, \ldots, n\}$ is a finite collection of open sets. Then we will prove that $A = \bigcap_{i=1}^{n} A_i$ is an open set.

Let $x \in A = \bigcap_{i=1}^{n} A_i$. Then we will show that there exists an open sphere containing x and contained in A.

Since $x \in A = \bigcap_{i=1}^{n} A_i$

$$\Rightarrow \quad x \in A_i, \quad \text{for all } i \in \wedge.$$

Since each A_i is an open set, by the definition of open sets, there exists an open sphere $S(x, r_i)$ such that

$$x \in S(x, r_i) \subseteq A_i \quad \text{for all } i \in \wedge.$$

Let $r = \min\{r_1, r_2, \ldots, r_n\}$. Then

$$\Rightarrow \quad x \in S(x,r) \subseteq S(x, r_i) \subseteq A_i \quad \text{for all } i \in \wedge.$$

$$\Rightarrow \quad x \in S(x,r) \subseteq A_i \quad \text{for all } i \in \wedge.$$

$$\Rightarrow \quad x \in S(x,r) \subseteq A = \bigcap_{i=1}^{n} A_i.$$

Since x is an arbitrary point, each point of A is an interior point and, therefore, $A = \bigcap_{i=1}^{n} A_i$ is an open set.

Remark. The intersection of an arbitrary number of open sets need not be an open set. Consider the usual metric d on $\mathbf{R}$ defined by

$$d(x, y) = |x - y|, \quad \forall x, y \in \mathbf{R}.$$

In usual metric space $(\mathbf{R}, d)$ open intervals are open sets. Consider

$$A_n = \left(-\frac{1}{n}, \frac{1}{n}\right), \quad n \in \mathbf{N}.$$

Each A_n is an open set with respect to the usual metric and $\bigcap_{n=1}^{\infty} A_n = \{0\}$. But $\{0\}$ is not an open set, because there does not exist any open sphere which contains and is contained in $\{0\}$. That is, there does not exist any $r > 0$ such that $S(0, r) \subseteq \{0\}$, since $S(0, r)$ is an infinite set and $\{0\}$ is a finite set.

Theorem 4. *In a metric space, a set is closed if and only if its complement is open.*

Proof. Let (X, d) be a metric space and let A be a closed set with respect to the metric d. Then we will show that its complement $A' = X - A = B$ is an open set. To show that B is open set, we need to show that each of its points is an interior point. Let $x \in B$ be an arbitrary point. Then we will show that there exists an open sphere centered at x and contained in B. Since $x \in B$, so $x \notin A$. Since A is a closed set, x cannot be its limit point. Then, for some $r > 0$, there exists an open sphere $S(x, r)$ such that

$$A \cap S(x, r) = \Phi,$$

$$\Rightarrow S(x, r) \subseteq B.$$

i.e. $x \in S(x, r) \subseteq B$.

So, x is an interior point of the set B. Since x is an arbitrary point, each point of the set B is an interior point. Hence, B is an open set.

Conversely, let $A' = X - A = B$ be an open set. Then we will show that A is a closed set. To show that A is a closed set, we need to show that A has each of its limit points. That is, no point outside from A is a limit point of A. Let, if possible, the point $x \notin A$ be a limit point of A. Since $x \notin A$, therefore, $x \in B$ and B is an open set, so there exists an open sphere $S(x, r)$ such that

$$x \in S(x, r) \subseteq B$$
$$\Rightarrow A \cap S(x, r) = \Phi,$$

which contradicts the fact that x is a limit point of A. So, a point outside A cannot be a limit point of A. Therefore, A is a closed set.

Theorem 5. *In a metric space, every closed sphere is a closed set.*

Proof. Let (X, d) be a metric space and let $S[a, r]$ be a closed sphere. Then we will show that $S[a, r]$ is a closed set. For this, we need to show that its complement $(S[a, r])' = X - S[a, r]$ is an open set. So, we will prove that each point in $(S[a, r])'$ is an interior point.

Let $y \in (S[a, r])'$ be an arbitrary point. Then we will show that y is an interior point of $(S[a, r])'$. For this, we need to show that there exists an open sphere centered at y and contained in $(S[a, r])'$.

Since $y \in (S[a, r])'$ by the definition of closed sphere, we have

$$d(y, a) > r.$$
$$\Rightarrow \quad d(y, a) - r > 0.$$

Let

$$r_1 = d(y, a) - r > 0. \tag{12.18}$$

Now, we will show that the open sphere $S(y, r_1)$ is contained in $(S[a, r])'$.

Let $x \in S(y, r_1)$ be an arbitrary point. Then we will show that $x \in (S[a, r])'$ for which we will prove that $d(x, a) > r$.

Since $x \in S(y, r_1)$

$$\Rightarrow \quad d(x, y) < r_1, \tag{12.19}$$

(X, d) is a metric space. Using the triangle inequality for the points y, a and x, we have

$$d(y, a) \leq d(y, x) + d(x, a),$$

Using Eq. (12.19) in the above inequality, we get

$$d(y, a) < r_1 + d(x, a).$$

Using Eq. (12.18), we have

$$d(y, a) < d(y, a) - r + d(x, a).$$
$$\Rightarrow d(x, a) > r.$$
$$\Rightarrow x \notin S[a, r].$$
$$\Rightarrow x \in (S[a, r])'.$$

Since x is an arbitrary point, each point in $S(y, r_1)$ is also in $(S[a, r])'$.

$$\Rightarrow y \in S(y, r_1) \subseteq (S[a, r])'.$$

So, y is an interior point. Since y is an arbitrary point, each point in $(S[a, r])'$ is an interior point. Therefore, $(S[a, r])'$ is an open set, and hence, $S[a, r]$ is a closed set.

Theorem 6. *In metric space, the intersection of an arbitrary collection of closed sets is a closed set.*

Proof. Let (X, d) be a metric space and let $\{A_\lambda\}_{\lambda \in \wedge}$ be an arbitrary collection of closed sets. Then we will prove that $A = \bigcap_{\lambda \in \wedge} A_\lambda$ is a closed set. To show that A is a closed set, we need to prove that its complement is open, that is, A' is open. Using De Morgan's law, we can write

$$A' = \bigcup_{\lambda \in \wedge} A'_\lambda, \qquad (12.20)$$

Since $\{A_\lambda\}_{\lambda \in \wedge}$ is an arbitrary collection of closed sets, A'_λ is an open set $\forall \lambda \in \wedge$. In Eq. (12.20), A' is the union of arbitrary collections of open sets, therefore, it is an open set. Hence, A is closed set.

Theorem 7. *In a metric space, the union of a finite number of closed sets is a closed set.*

Proof. Let (X, d) be a metric space and let $\{A_\lambda\}_{\lambda \in \wedge}$, where $\wedge = \{1, 2, \ldots, n\}$ be a finite collection of closed sets. Then we will prove that $A = \bigcup_{i=1}^{n} A_i$ is a closed set.

By De Morgan's law, we can write

$$\left(\bigcup_{i=1}^{n} A_i\right)' = \bigcap_{i=1}^{n} A_i'. \tag{12.21}$$

Since each A_i is a closed set, each A_i' is an open set. Since the right-hand side in Eq. (12.21) is a finite intersection of open sets, it is open. So $(\bigcup_{i=1}^{n} A_i)'$ is an open set. Hence, $\bigcup_{i=1}^{n} A_i$ is a closed set.

Remark. The union of an arbitrary number of closed sets need not be closed. Consider the usual metric d on $\mathbf{R}$, defined by

$$d(x, y) = |x - y|, \quad \forall x, y \in \mathbf{R}.$$

In the usual metric space $(\mathbf{R}, d)$, closed intervals are closed sets. Consider

$$A_n = \left[\frac{1}{n}, 1\right], \quad n \in \mathbf{N}.$$

Then each A_n is a closed set with respect to the usual metric. But their union
$A = \bigcup_{n=1}^{\infty} A_n = (0, 1]$, which is not a closed set, because is the limit point of the set A but $0 \notin (0, 1]$.

Subspace. Let (X, d) be a metric space and let $Y \subseteq X$. Then the metric space (Y, d) is a subspace of (X, d).

Example. Consider the usual metric d on $\mathbf{R}$, defined by

$$d(x, y) = |x - y|, \quad \forall x, y \in \mathbf{R}.$$

If $(\mathbf{R}, d)$ is a metric space and $\mathbf{Q} \subset \mathbf{R}$, then $(\mathbf{Q}, d)$ is a subspace of $(\mathbf{R}, d)$.

12.3 Bounded Sequence

A sequence $\langle x_n \rangle$ in a metric space (X, d) is said to be bounded above if there exists a real number M such that $x_n \leq M$ ($\forall n \in \mathbf{N}$).

A sequence $\langle x_n \rangle$ in a metric space (X, d) is said to be bounded below if there exists a real number m such that $m \leq x_n$ ($\forall n \in \mathbf{N}$).

A sequence $\langle x_n \rangle$ in a metric space (X, d) is said to be bounded if it is bounded above as well as bounded below. That is, there exist two real numbers m and M such that $m \leq x_n \leq M$ ($\forall n \in \mathbf{N}$). We can also write $|x_n| \leq M$ ($\forall n \in \mathbf{N}$).

Simply, we can say that a sequence $\langle x_n \rangle$ in a metric space (X, d) is bounded if its range is bounded.

Convergence of Sequences. Let (X, d) be a metric space. A sequence $\langle x_n \rangle$ is said to have a limit l if, for each $\varepsilon > 0$, there exists a natural number m depending on ε such that

$$d(x_n, l) < \varepsilon, \forall n \geq m.$$

A sequence $\langle x_n \rangle$ is said to be convergent if it has a finite limit l.

A sequence $\langle x_n \rangle$ is said to be convergent to a finite limit l if only a finite number of terms of the sequence lie outside the open sphere $S(l, \varepsilon)$.

Theorem 1. *In a metric space, every convergent sequence is bounded, but the converse is not true.*

Proof. Let (X, d) be a metric space and let $\langle x_n \rangle$ be a convergent sequence which converges to l. Then, for $\varepsilon > 0$, there exists a natural number m such that

$$d(x_n, l) < \varepsilon, \forall n \geq m.$$

Let $\varepsilon = 1$. Then

$$d(x_n, l) < 1, \forall n \geq m.$$

Therefore, only a finite number of terms of the sequence lie outside the open sphere $S(l, 1)$.

Let $M = \max\{d(x_1, l), d(x_2, l), \ldots, d(x_{m-1}, l), 1\}$. Then

$$d(x_n, l) < M, \quad \forall n \in \mathbf{N}. \tag{12.22}$$

Now, using the triangle inequality, we have

$$d(x_n, x_m) \leq d(x_n, l) + d(l, x_m).$$

Using Eq. (12.22), we can write

$$d(x_n, x_m) < 2M, \quad \forall n \in \mathbf{N}.$$

Hence, the sequence $\langle x_n \rangle$ is bounded.

To show that the converse is not true, we give an example of a sequence which is bounded, but not convergent. Consider the usual metric d on $\mathbf{R}$, defined by

$$d(x, y) = |x - y|, \quad \forall x, y \in \mathbf{R}.$$

Consider the sequence $\langle (-1)^n \rangle = \langle -1, 1, -1, 1, \ldots \rangle$. Then $-1 \leq x_n \leq 1 \; \forall n \in \mathbf{N}$. So, the sequence $\langle (-1)^n \rangle$ is bounded. If possible, let $\lim_{n \to \infty} x_n = l$. Then, for $\varepsilon = \frac{1}{3}$, there exists a natural number m such that

$$|x_n - l| < \frac{1}{3}, \quad \forall n \geq m.$$

$$\Longleftrightarrow |(-1)^n - l| < \frac{1}{3}, \quad \forall n \geq m.$$

In particular, taking $n = 2m$ and $n = 2m + 1$, we have

$$|(-1)^{2m} - l| < \frac{1}{3} \quad \text{and} \quad |(-1)^{2m+1} - l| < \frac{1}{3}.$$

$$\Rightarrow |1 - l| < \frac{1}{3} \quad \text{and} \quad |-1 - l| < \frac{1}{3}.$$

$$\Rightarrow |1 - l| < \frac{1}{3} \quad \text{and} \quad |1 + l| < \frac{1}{3}.$$

Now, $2 = |1 - l + 1 + l| \leq |1 - l| + |1 + l| < \frac{1}{3} + \frac{1}{3} = \frac{2}{3}$.

$$\Rightarrow 2 < \frac{2}{3},$$

which is absurd. So, our assumption that the sequence $\langle (-1)^n \rangle$ converges to l is false. Hence, the sequence $\langle (-1)^n \rangle$ is not convergent.

Theorem 2. *In a metric space, a sequence cannot converge to more than one limit.*

Proof. Let (X, d) be a metric space. If possible, let $\langle x_n \rangle$ be a sequence which converges to two different limits l and l'.

Let the sequence $\langle x_n \rangle$ converge to l. Then, for $\varepsilon > 0$, there exists a natural number n_1 such that

$$d(x_n, l) < \frac{\varepsilon}{2} \quad \text{for all } n \geq n_1. \tag{12.23}$$

Let the sequence $\langle x_n \rangle$ converge to l'. Then, for $\varepsilon > 0$, there exists a natural number n_2 such that

$$d(x_n, l') < \frac{\varepsilon}{2} \quad \text{for all } n \geq n_2. \tag{12.24}$$

Let $m = \max\{n_1, n_2\}$. Then, from Eqs. (12.23) and (12.24), we can write

$$d(x_n, l) < \frac{\varepsilon}{2} \quad \text{and} \quad d(x_n, l') < \frac{\varepsilon}{2} \quad \text{for all } n \geq m. \tag{12.25}$$

Let us assume that $\varepsilon = \frac{1}{2}d(l, l') > 0$, because $l \neq l'$. Then $d(l, l') \neq 0$.

Now, from the triangle inequality for points l, l' and x_n, we can write

$$d(l, l') \leq d(l, x_n) + d(x_n, l'). \tag{12.26}$$

Using Eq. (12.25) in Eq. (12.26), we get

$$d(l, l') < \varepsilon + \varepsilon = 2\varepsilon,$$

$$\Rightarrow d(l, l') < 2\frac{1}{2}d(l, l')$$

$$\Rightarrow d(l, l') < d(l, l').$$

This is absurd. So, our assumption that a sequence converges to more than one limit in a metric space (X, d) is false. Hence, in a metric space, a sequence cannot converge to more than one limit.

Cauchy sequences. Let (X, d) be a metric space. A sequence $\langle x_n \rangle$ is said to be a Cauchy sequence if, for a given $\varepsilon > 0$, there exists a natural number m such that

$$d(x_n, x_m) < \varepsilon \quad \text{whenever } n \geq m.$$

Equivalently, a sequence $\langle x_n \rangle$ is said to be a Cauchy sequence if, for a given $\varepsilon > 0$, there exists a natural number p such that

$$d(x_n, x_m) < \varepsilon \quad \text{whenever } n, m \geq p.$$

Example 1. Consider the usual metric d on $\mathbf{R}$ defined by

$$d(x, y) = |x - y|, \quad \forall x, y \in \mathbf{R}.$$

Then the sequence $\langle \frac{1}{n} \rangle$ is a Cauchy sequence in $(\mathbf{R}, d)$.

Let $\varepsilon > 0$ be given and $x_n = \frac{1}{n}$. Then, for $n > m$,

$$d(x_n, x_m) = |x_n - x_m| = \left| \frac{1}{n} - \frac{1}{m} \right|$$

$$\leq \left| \frac{1}{n} \right| + \left| \frac{1}{m} \right| = \frac{1}{n} + \frac{1}{m}.$$

By the Archimedean property, we can find a natural number p such that $p > \frac{2}{\varepsilon}$. Then, for $n, m \geq p$, we have

$$\frac{1}{n}\frac{1}{m} \leq \frac{1}{p} < \frac{\varepsilon}{2},$$

$$\Rightarrow \frac{1}{n} < \frac{\varepsilon}{2} \quad \text{and} \quad \frac{1}{m} < \frac{\varepsilon}{2}.$$

$$\Rightarrow |x_n - x_m| \leq \frac{1}{n} + \frac{1}{m} < \frac{\varepsilon}{2} + \frac{\varepsilon}{2} = \varepsilon.$$

So, for $n, m \geq p$, we have

$$d(x_n, x_m) = |x_n - x_m| < \varepsilon$$

$$\Rightarrow d(x_n, x_m) < \varepsilon.$$

Therefore, $\langle \frac{1}{n} \rangle$ is a Cauchy sequence in $(\mathbf{R}, d)$.

Example 2. Consider the usual metric d on $\mathbf{R}$ defined by

$$d(x, y) = |x - y|, \quad \forall x, y \in \mathbf{R}.$$

Then the sequence $\langle \frac{(-1)^n}{n} \rangle$ is a Cauchy sequence in $(\mathbf{R}, d)$.

Let $\varepsilon > 0$ be given and $x_n = \frac{(-1)^n}{n}$. Then, for $n > m$, we have

$$d(x_n, x_m) = |x_n - x_m| = \left| \frac{(-1)^n}{n} - \frac{(-1)^m}{m} \right|$$

$$\leq \left| \frac{(-1)^n}{n} \right| + \left| \frac{(-1)^m}{m} \right| = \frac{1}{n} + \frac{1}{m},$$

By the Archimedean property, we can find a natural number p such that $p > \frac{2}{\varepsilon}$. Then, for $n, m \geq p$, we have $|x_n - x_m| \leq \frac{1}{n} + \frac{1}{m} < \frac{\varepsilon}{2} + \frac{\varepsilon}{2} = \varepsilon$.

So, for $n, m \geq p$, we have

$$d(x_n, x_m) = |x_n - x_m| < \varepsilon.$$

$$\Rightarrow d(x_n, x_m) < \varepsilon.$$

Therefore, $\langle \frac{(-1)^n}{n} \rangle$ is a Cauchy sequence in $(\mathbf{R}, d)$.

Example 3. Consider $d : X \times X \to \mathbf{R}$ defined by

$$d(x, y) = \begin{cases} 1, & \text{if } x \neq y, \\ 0 & \text{if } x = y, \end{cases} \quad \forall x, y \in X.$$

Then (X, d) is a discrete metric space. In a discrete metric space, constant sequences are Cauchy sequences.

Let $\langle x_n \rangle$ be a constant sequence in (X, d). Then

$$d(x_n, x_m) = 0. \quad \forall n \in \mathbf{N} \text{ (because } x_n = x_m \quad \forall n \in \mathbf{N}).$$

Then, for each $\varepsilon > 0$, we can write

$$d(x_n, x_m) < \varepsilon, \quad \forall n, m \in \mathbf{N}.$$

Therefore, $\langle x_n \rangle$ is a Cauchy sequence in (X, d).

Theorem. *In a metric space, every convergent sequence is a Cauchy sequence, but a Cauchy sequence need not be convergent.*

Proof. Let (X, d) be a metric space and let $\langle x_n \rangle$ be a convergent sequence which converges to a limit l. Then we will prove that $\langle x_n \rangle$ is a Cauchy sequence. Let $\varepsilon > 0$ be given. Then there exists a natural number m such that

$$d(x_n, l) < \frac{\varepsilon}{2}, \quad \forall n \geq m. \tag{12.27}$$

In particular, when $n = m$, we can write

$$d(x_m, l) < \frac{\varepsilon}{2}. \tag{12.28}$$

Now, from the triangle inequality, we can write

$$d(x_n, x_m) \leq d(x_n, l) + d(l, x_m). \tag{12.29}$$

So, for $\varepsilon > 0$ and $n \geq m$, from Eqs. (12.27), (12.28), and (12.29), we have

$$d(x_n, x_m) < \frac{\varepsilon}{2} + \frac{\varepsilon}{2} = \varepsilon$$

$$\Rightarrow d(x_n, x_m) < \varepsilon, \quad \forall n \geq m.$$

So, $< x_n >$ is a Cauchy sequence in (X, d).

Therefore, in a metric space, every convergent sequence is a Cauchy sequence.

To show that the converse is not true, we will give an example of a Cauchy sequence which is not convergent.

Consider the usual metric d on $X = (0, 1]$ defined by

$$d(x, y) = |x - y|, \quad \forall x, y \in X.$$

Then the sequence $\langle \frac{1}{n} \rangle$ is a Cauchy sequence in (X, d). We have already proved it. The sequence $\langle \frac{1}{n} \rangle$ converges to but $0 \notin X = (0, 1]$.

Therefore, the sequence $\langle \frac{1}{n} \rangle$ is a Cauchy sequence in (X, d) but it does not converge in (X, d).

- A convergent sequence is a Cauchy sequence in any general metric space, but the converse is not true. Now, the question arises as to under what condition does a Cauchy sequence converge in a metric space. This gives arise to the concept of a new metric space called complete metric space.

12.4   Complete Metric Space

A metric space (X, d) is said to be complete if every Cauchy sequence in (X, d) converges in (X, d). That is, if $\langle x_n \rangle$ is a Cauchy sequence in (X, d), then $x_n \to x$ and $x \in X$.

Theorem 1. *Let (Y, d) be a subspace of a complete metric space (X, d). Then (Y, d) is complete if and only if Y is closed.*

Proof. Let (X, d) be a complete metric space and let (Y, d) be a subspace of (X, d).

Let (Y, d) be a complete metric space. Then we will show that Y is closed for this, we will show that it has all its limit points. Let $x \in X$ be a limit point of Y. Then we will prove that $x \in Y$.

Since $x \in X$ is a limit point of Y by the definition of a limit point, we can always construct a sequence $\langle x_n \rangle$ in Y which converges to x. Further, since every convergent sequence is a Cauchy sequence, $\langle x_n \rangle$ is a Cauchy sequence in Y.

Since (Y, d) is a complete metric space, by the definition of a complete metric space, a sequence in Y converges in Y.

$$\Rightarrow \ x \in Y.$$

Since x is an arbitrary limit point, Y has all of its limit points. So Y is closed.

To show that the converse is true, let Y be a closed subspace of (X, d). Then we will show that (Y, d) is a complete metric space.

Let $\langle x_n \rangle$ be a Cauchy sequence in Y.

Since $Y \subseteq X$, so $\langle x_n \rangle$ is a Cauchy sequence in X. Since (X, d) is a complete metric space, the sequence $x_n \to x$ and $x \in X$. Since x is a limit, it is also a limit point.

So x is a limit point of the sequence $\langle x_n \rangle$ in Y. Since Y is closed, it has all of its limit points.

$$\Rightarrow \ x \in Y.$$

Since $\langle x_n \rangle$ is an arbitrary Cauchy sequence in Y and since we have proved that it converges in Y, (Y, d) is a complete metric space.

Theorem 2. *Show that $(\mathbf{R}, d)$ is a complete metric space with respect to the usual metric given by*

$$d(x, y) = |x - y|, \quad \forall x, y \in \mathbf{R}.$$

Proof. Let $\langle x_n \rangle$ be an arbitrary Cauchy sequence in $(\mathbf{R}, d)$. Choose $\varepsilon = \frac{1}{2^{k+1}}$ and define a sequence $\langle n_k \rangle$ of natural numbers by setting n_{k+1} as the smallest natural number $\geq n_k$ such that

$$d(x_n, x_m) = |x_n - x_m| < \frac{1}{2^{k+1}} \quad \text{whenever } n, m \geq n_k. \qquad (12.30)$$

Let us define a sequence $\{I_k\}$ of closed intervals by

$$I_k = \left[x_{n_k} - \frac{1}{2^k}, x_{n_k} + \frac{1}{2^k} \right].$$

Then $I_{k+1} = \left[x_{n_{k+1}} - \frac{1}{2^{k+1}}, x_{n_{k+1}} + \frac{1}{2^{k+1}} \right]$. Also, from Eq. (12.30), we have

$$|x_{n_k} - x_{n_{k+1}}| < \frac{1}{2^{k+1}}.$$

$$\Rightarrow I_{k+1} \subset I_k.$$

Further, we have $(I_k) = \frac{2}{2^{k+1}} = \frac{1}{2^k}$.

Then $\lim_{k \to \infty} l(I_k) = 0$.

So, the sequence $\{I_k\}$ is a nested sequence of closed intervals with $\lim_{k \to \infty} l(I_k) = 0$. Then, by Cantor's intersection theorem, we have

$$\bigcap_{k=1}^{\infty} I_k = \{x\}.$$

Since x is in intersection, therefore, $x \in I_k$. From the construction of I_k, it is clear that

$$|x - x_{n_k}| < \frac{1}{2^k}. \quad \forall n_k \in \mathbf{N}. \tag{12.31}$$

Taking $m = n_k$ in Eq. (12.30), we get

$$|x_n - x_{n_k}| < \frac{1}{2^{k+1}}. \tag{12.32}$$

Now, for $\forall \; n \geq n_k$, we have

$$|x_n - x| = |x_n - x_{n_k} + x_{n_k} - x|,$$
$$\leq |x_n - x_{n_k}| + |x_{n_k} - x|.$$

Using Eqs. (12.31) and (12.32), we get

$$|x_n - x| < \frac{1}{2^k} + \frac{1}{2^{k+1}} < \frac{1}{2^k} + \frac{1}{2^k}.$$
$$\Rightarrow |x_n - x| < \frac{1}{2^{k-1}}. \quad \forall n \geq n_k. \tag{12.33}$$

For a given $\varepsilon > 0$, by the Archimedean property, we can choose a natural number n_k such that $\frac{1}{2^{k-1}} < \varepsilon$. Then, from Eq. (12.33), we can write

$$|x_n - x| < \varepsilon. \quad \forall n \geq n_k.$$
$$\Rightarrow x_n \to x.$$

Since $\langle x_n \rangle$ is an arbitrary Cauchy sequence in $(\mathbf{R}, d)$ and since we have proved it converges in $(\mathbf{R}, d)$, so $(\mathbf{R}, d)$ is a complete metric space.

12.5 Cantor's Intersection Theorem

Theorem. *Let* (X, d) *be a complete metric space and let* $\{A_n\}$ *be a nested sequence of non-empty closed intervals in* (X, d) *such that* $\delta(A_n) \to 0$ *as* $n \to \infty$. *Then* $\bigcap_{n=1}^{\infty} A_n \neq \varphi$ *and* $\bigcap_{n=1}^{\infty} A_n$ *has exactly one point.*

Proof. Let (X, d) be a complete metric space and let $\{A_n\}$ be a nested sequence of non-empty closed intervals in (X, d). Since $\{A_n\}$ is a nested sequence, we have

$$A_1 \supseteq A_2 \supseteq A_3 \supseteq \cdots A_n \supseteq \cdots$$

First, we will construct a Cauchy sequence in X. Let $\langle x_n \rangle$ be a sequence such that $x_n \in A_n$, $\forall n \in \mathbf{N}$. Since $\{A_n\}$ are nested,

$$x_n \in A_p \quad \forall n \geq p.$$

Since $\delta(A_n) \to 0$ as $n \to \infty$, for $\varepsilon > 0$, there exists a natural number p such that

$$\delta(A_n) < \varepsilon, \quad \forall n \geq p.$$

In particular, we have

$$\delta(A_p) < \varepsilon.$$

Since $x_n, x_m \in A_p$ $\forall m, n \geq p$, we have

$$d(x_n, x_m) < \delta(A_p) < \varepsilon, \quad \forall mn \geq p. \tag{12.34}$$

So $\langle x_n \rangle$ is a Cauchy sequence in X. Since (X, d) is a complete metric space, $x_n \to x$ and $x \in X$.

Now, we will claim that $x \in \bigcap_{n=1}^{\infty} A_n$.

Since $x_n \in A_n, ; x_{n+1} \in A_{n+1}; x_{n+2} \in A_{n+2}; \ldots$ the subsequence $\{x_n, x_{n+1}, \ldots\}$ will be in A_n because $\{A_n\}$ are nested. It is a Cauchy sequence and A_n is a closed subspace of X. So, it is a complete metric space. Then, by the definition of completeness, $x \in A_n$. A similar process can be applied for each A_n.

Therefore, $x \in A_n \forall n \in \mathbf{N}$.

$$\Rightarrow x \in \bigcap_{n=1}^{\infty} A_n.$$

To show the uniqueness, let x and $y \in \bigcap_{n=1}^{\infty} A_n$. Then we will show that $x = y$. Now,

$$x \text{ and } y \in A_n, \quad \forall n \in \mathbf{N}.$$
$$0 \leq d(x, y) \leq \delta(A_n) \to 0 \quad \text{as } n \to \infty.$$
$$\Rightarrow d(x, y) = 0$$
$$\Rightarrow x = y.$$

The proof is thus completed.

12.6   Continuous Functions on Metric Spaces

Let (X, d_1) and (Y, d_2) be two metric spaces. A function $f : X \to Y$ is said to be continuous at a point b of X if, for a given $\varepsilon > 0$, there exists a $\delta > 0$ such that

$$d_2(f(x), f(b)) < \varepsilon \quad \text{whenever } d_1(x, b) < \delta.$$

We can equivalently write it as "for each open sphere $S(f(b), \varepsilon)$ there exists an open sphere $S(b, \delta)$ such that $f(S(b, \delta)) \subseteq S(f(b)\varepsilon)$".

If f is continuous at every point of the domain X then it is said to be continuous on X.

Example 1. Let (X, d_1) and (Y, d_2) be any two metric spaces. A constant function $f : X \to Y$ is continuous on X.

Example 2. If (X, d) is a discrete metric space, then every function $f : X \to Y$ is continuous on X.

For any $b \in X$ if we choose $\delta < 1$, then $S(b, \delta) = \{b\}$ and so $f(S(b, \delta)) = \{f(b)\} \subseteq S(f(b), \varepsilon)$ for every $\epsilon > 0$.

12.7   Uniform Continuous Functions on Metric Spaces

Let (X, d_1) and (Y, d_2) be two metric spaces. A function $f : X \to Y$ is said to be uniformly continuous on X if, for given $\varepsilon > 0$, there exists a δ depending only on ε such that

$$d_2(f(x), f(y)) < \varepsilon \quad \text{whenever } d_1(x, y) < \delta, \ \forall x, y \in X.$$

Theorem 1. *Let (X, d_1) and (Y, d_2) be any two metric spaces. A function $f : X \to Y$ is continuous on X if and only if, for each open set $G \subset Y$, $f^{-1}(G)$ is an open subset of X.*

Proof. Let (X, d_1) and (Y, d_2) be any two metric spaces and let $f : X \to Y$ be a continuous function. Let G be an open subset of Y. Then we will show that $f^{-1}(G)$ is an open subset in X. If $f^{-1}(G) = \varphi$, then it is open. If $f^{-1}(G) \neq \varphi$, let b be any point of $f^{-1}(G)$ so that $f(b) \in G$. Since G is an open set containing $f(b)$, there exists an open sphere $S(f(b), \varepsilon)$ for some $\varepsilon > 0$ such that

$$S(f(b), \varepsilon) \subseteq G. \tag{12.35}$$

Since f is continuous at b, we can find a $\delta > 0$ such that

$$f(S(b, \delta)) \subseteq S(f(b), \varepsilon). \tag{12.36}$$

From (12.35) and (12.36), we have

$$f(S(b, \delta)) \subseteq G \quad \text{so that } S(b, \delta) \subseteq f^{-1}(G),$$

showing that each point of $f^{-1}(G)$ is the center of an open sphere contained in it and hence $f^{-1}(G)$ is an open set in X.

Conversely, we assume that the inverse image under f of every open subset of Y is an open subset of X. Then we will show that f is continuous. For this it is sufficient to show that it is continuous at an arbitrary point $b \in X$. Let $\varepsilon > 0$. Then $S(f(b), \varepsilon)$ is open in $Y \Rightarrow f^{-1}(S(f(b), \varepsilon))$ is open in X.

Hence, there exists $\delta > 0$ such that

$$S(b, \delta) \subseteq f^{-1}(S(f(b), \varepsilon)) \quad \text{so that } f(S(b, \delta)) \subseteq S(f(b)\varepsilon).$$

$\Rightarrow f$ is continuous at b.

Since b is an arbitrary point of X, so f is continuous on X.

Open cover. Let (X, d) be a metric space and let F be a subset of X. A family $A = \{G_i : i \in \Lambda\}$ of open subsets of X is said to be an open cover of F if

$$F \subset \cup\{G_i : i \in \Lambda\},$$

where Λ is an index set.

12.8 Compact Metric Space

A subset F of a metric space (X, d) is said to be compact if every open cover of F admits of a finite subcover, that is, each family $\{G_i : i \in \Lambda\}$ of open subsets of X for which $F \subset \cup\{G_i : i \in \Lambda\}$, there exists a finite subcover $\{G_{i_1}, G_{i_2}, \ldots, G_{i_n}\}$ such that $F \subset \cup\{G_{i_j} : j = 1, 2 \ldots n\}$.

A metric space (X, d) is said to be compact if every open cover of X admits of a finite subcover, that is, for each family $\{G_i : i \in \Lambda\}$ of open subsets of X for which $X = \cup\{G_i : i \in \Lambda\}$, there exists a finite subcover $\{G_{i_1}, G_{i_2}, \ldots, G_{i_n}\}$ such that $X = \cup\{G_{i_j} : j = 1, 2, \ldots, n\}$.

Example 1. In the usual metric, every closed interval is compact.

Example 2. The discrete space (X, d), where X is a finite set, is compact.

Example 3. The discrete space (X, d), where X is an infinite set, is not compact.

For each $x \in X$, $\{x\}$ is open in X.

Also $\bigcup_{x \in X} \{x\} = X$.

Therefore, $\{\{x\} : x \in X\}$ is an open cover of X. This open cover has no finite subcover because X is an infinite set.

Theorem 1. *Every closed subset of a compact metric space is compact.*

Proof. Let (X, d) be a compact metric space and let F be a closed subset of X. Then we will show that F is a compact set. Let $A = \{G_i : i \in \Lambda\}$ be an open cover of F. Then $A^* = \{G_i : i \in \Lambda\} \cup (X - F)$

is an open cover of X. Since X is compact, it has a finite subcover, say $G_{i_1} G_{i_2}, \ldots, G_{i_n}, X - F$, so that

$$G_{i_1} \cup G_{i_2} \cup \ldots, G_{i_n} \cup X - F = X,$$

and so $F \subset \cup \{G_{i_j} : j = 1, 2, \ldots, n\}$.

$\Rightarrow \{G_{i_j} : j = 1, 2 \ldots n\}$ is a finite subcover of F. Therefore, F is compact.

Theorem 2. *A continuous image of a compact metric space is compact.*

Proof. Let f be a continuous function from a compact metric space X into a metric space Y. Then we will show that $f(X)$ is a compact set. Let $A = \{G_i : i \in \Lambda\}$ be an open cover of $f(X)$,

$$f(X) \subset \cup \{G_i : i \in \Lambda\}. \tag{12.37}$$

Then we will show the it will admit of a finite subcover. Since f is a continuous function, $f^{-1}(G_i)$ is an open subset of X for every open set G_i in A. From Eq. (12.37), we have

$$X \subset f^{-1}(f(X)) \subset f^{-1}\{\cup G_i : i \in \Lambda\},$$

$$\Rightarrow X \subset \cup \{f^{-1}(G_i) : i \in \Lambda\},$$

which shows that $\{f^{-1}(G_i) : i \in \Lambda\}$ is an open cover of X. Since X is compact, it will admit of a finite subcover so that

$$X \subseteq \cup \{f^{-1}(G_{i_j}) : j = 1, 2 \ldots n\}$$

$$\Rightarrow f(X) \subset f(\cup \{f^{-1}(G_{i_j}) : j = 1, 2 \ldots n\})$$

$$\Rightarrow f(X) \subset \cup \{G_{i_j} : j = 1, 2 \ldots n\}.$$

Thus, $f(X)$ admits of a finite subcover, so $f(X)$ is compact.

Hence, a continuous image of a compact metric space is compact.

Separated sets. Let (X, d) be a metric space. Two non-empty subsets G and H are said to be separated if G and H are disjoint and neither G nor H contains any limit point of the other. That is,

$$\bar{G} \cap H = \varphi \quad \text{and} \quad G \cap \bar{H} = \varphi.$$

Example 1. Consider $G = (0, 2)$ and $H = (2, 3)$. Then G and H are disjoint. Also $\bar{G} = [0, 2]$ and $\bar{H} = [2, 3]$ so that $\bar{G} \cap H = \varphi$ and $G \cap \bar{H} = \varphi$. Hence, G and H are separated.

Example 2. Consider $G = (0, 2)$ and $H = [2, 3)$. Then G and H are disjoint. But $\bar{G} = [0, 2]$ and $\bar{H} = [2, 3]$, so that $\bar{G} \cap H = \{2\}$ and $G \cap \bar{H} = \varphi$. Hence, G and H are disjoint, but not separated.

12.9   Connected and Disconnected Sets

A subset E of a metric space (X, d) is said to be connected if it cannot be expressed as the union of two non-empty separated sets. If E is not connected, then it is said to be disconnected. When E is disconnected, then we have two-empty subsets G and H of X such that

$$\bar{G} \cap H = \varphi \quad \text{and} \quad G \cap \bar{H} = \varphi \quad \text{and} \quad E = G \cup H.$$

Example. Any discrete metric space with more than one point is disconnected.

Theorem 1. *Let (X, d) be a metric space and let F be a connected subset of X such that $F \subset G \cup H$, where G and H are separated subsets of X. Then either $F \subset G$ or $F \subset H$.*

Proof.   Since G and H are separated subsets of X

$$\Rightarrow \quad \bar{G} \cap H = \varphi \quad \text{and} \quad G \cap \bar{H} = \varphi \qquad (12.38)$$

Again, since $F \subset G \cup H$, we have

$$F = F \cap (G \cup H) = (F \cap G) \cup (F \cap H). \qquad (12.39)$$

We now show that at least one of the sets $F \cap G$ and $F \cap H$ is empty. If possible, let neither $F \cap G \neq \varphi$ nor $F \cap H \neq \varphi$. Then

$$(F \cap G) \cap \overline{(F \cap H)} \subset (F \cap G) \cap (\bar{F} \cap \bar{H}) = (F \cap \bar{F}) \cap (G \cap \bar{H}).$$

Using Eq. (1), we get

$$(F \cap G) \cap \overline{(F \cap H)} \subset (F \cap \bar{F}) \cap \varphi = \varphi.$$
$$\Rightarrow (F \cap G) \cap \overline{(F \cap H)} \subset \varphi. \qquad (12.40)$$

Again, we have

$$\overline{(F \cap G)} \cap (F \cap H) \subset (\bar{F} \cap \bar{G}) \cap (F \cap H) = (\bar{F} \cap F) \cap (\bar{G} \cap H).$$

Using Eq. (12.38) again, we have

$$\overline{(F \cap G)} \cap (F \cap H) \subset (\bar{F} \cap F) \cap \varphi = \varphi.$$

$$\Rightarrow \overline{(F \cap G)} \cap (F \cap H) \subset \varphi. \tag{12.41}$$

From Eqs. (12.39), (12.40), and (12.41), we can see that F can be expressed as the union of two non-empty disjoint separated sets, which contradicts the fact that F is a connected set. Hence, at least one of the sets $F \cap G$ and $F \cap H$ is empty.

If $F \cap G = \varphi$, then, by Eq. (12.39),

$$F = \varphi \cup (F \cap H) = F \cap H$$

$$\Rightarrow F \subset H.$$

If $F \cap H = \varphi$, then, by Eq. (12.39),

$$F = (F \cap G) \cup \varphi = F \cap G.$$

$$\Rightarrow F \subset G.$$

Hence, either $F \subset G$ or $F \subset H$.

Theorem 2. *Let (X, d) be a metric space and let E be a connected subset of X. If F is a subset of X such that $E \subset F \subset \bar{E}$, then F is also connected. In particular, $\bar{E}$ is connected.*

Proof. We have to prove that F is connected. If possible, let F be disconnected. Then we have two-empty subsets G and H of X such that

$$\bar{G} \cap H = \varphi \quad \text{and} \quad G \cap \bar{H} = \varphi \quad \text{and} \quad F = G \cup H. \tag{12.42}$$

Given that

$$E \subset F \subset \bar{E}, \tag{12.43}$$

from Eqs. (12.42) and (12.43), we get

$$F = G \cup H \quad \text{and} \quad E \subset F \Rightarrow E \subset G \cup H. \tag{12.44}$$

Equation (12.44) shows that, if a connected set E is contained in the union of two separated sets G and H, then, by using the previous theorem, either $E \subset G$ or $E \subset H$.

If $E \subset G$, then $\bar{E} \subset \bar{G} \;\Rightarrow\; \bar{E} \cap H \subset \bar{G} \cap H \subset \varphi$, by Eq. (1),

$$\Rightarrow \bar{E} \cap H = \varphi. \tag{12.45}$$

From Eqs. (12.42) and (12.43), we see that

$$F = G \cup H \quad \text{and} \quad F \subset \bar{E} \;\Rightarrow\; G \cup H \subset \bar{E}.$$

$$\Rightarrow H \subset \bar{E} \;\Rightarrow\; \bar{E} \cap H = H. \tag{12.46}$$

From Eqs. (12.45) and (12.46), we have $H = \varphi$.

This contradicts the fact that H is non-empty. Hence, F must be connected.

Since $E \subseteq \bar{E} \subseteq \bar{E}$ and E is connected, by taking $F = \bar{E}$ and applying the above theorem, we see that $\bar{E}$ is connected.

Exercises

1. Let $d(x, y) = \min\{2, |x - y|\}$. Show that d is a metric.
2. Let (X, d) be a metric space if $d_1(x, y) = 5d(x, y)$. Then prove that (X, d_1) is also a metric space.
3. Let d be a function $d : \mathbf{R}^2 \times \mathbf{R}^2 \to \mathbf{R}$, defined by

$$d(x, y) = \max\{|x_1 - y_1|, |x_2 - y_2|\}, \quad \forall x = (x_1, x_2), y = (y_1, y_2) \in \mathbf{R}^2.$$

Then show that d is a metric on $\mathbf{R}^2$ and $(\mathbf{R}^2, d)$ is a metric space.

Bibliography

Apostol, T. M. *Mathematical Analysis*. Addison-Wesley (1974).

Bartle, R. G. *The Elements of Real Analysis*, 2nd edn. John Wiley & Sons (1976).

Bartle, R. G. and Sherbert, D. R. *Introduction to Real Analysis*. John Wiley & Sons (1982).

Burkill, J. C. *A First Course in Mathematical Analysis*. Cambridge University Press (1970).

Burkill, J. C. and Burkill, H. *A Second Course in Mathematical Analysis*. Cambridge University Press (1970).

Goffman, C. *Introduction to Real Analysis*, International edn. Harper & Row (1967).

Goldberg, R. R. *Methods of Real Analysis*. John Wiley & Sons (1976).

Malik, S. C. and Arora, S. *Mathematical Analysis*. New Age International Publisher, New Delhi (1982).

Narayan, S. *A Course of Mathematical Analysis*. S. Chand & Company, New Delhi (1958).

Royden, H. L. *Real Analysis*, 3rd edn. Macmillan, New York (1988).

Rudin, W. *Principles of Mathematical Analysis*, 3rd edn. McGraw-Hill, New York (1976).

Index

Printed in the United States
by Baker & Taylor Publisher Services